Advances in Experimental and Computational Rheology, Volume II

Advances in Experimental and Computational Rheology, Volume II

Editors

Maria Teresa Cidade
João Miguel Nóbrega

MDPI • Basel • Beijing • Wuhan • Barcelona • Belgrade • Manchester • Tokyo • Cluj • Tianjin

Editors
Maria Teresa Cidade
Universidade Nova de Lisboa
Portugal

João Miguel Nóbrega
University of Minho
Portugal

Editorial Office
MDPI
St. Alban-Anlage 66
4052 Basel, Switzerland

This is a reprint of articles from the Special Issue published online in the open access journal *Fluids* (ISSN 2311-5521) (available at: https://www.mdpi.com/journal/fluids/special_issues/rheology_II).

For citation purposes, cite each article independently as indicated on the article page online and as indicated below:

LastName, A.A.; LastName, B.B.; LastName, C.C. Article Title. *Journal Name* **Year**, *Article Number*, Page Range.

ISBN 978-3-03943-565-4 (Hbk)
ISBN 978-3-03943-566-1 (PDF)

© 2020 by the authors. Articles in this book are Open Access and distributed under the Creative Commons Attribution (CC BY) license, which allows users to download, copy and build upon published articles, as long as the author and publisher are properly credited, which ensures maximum dissemination and a wider impact of our publications.
The book as a whole is distributed by MDPI under the terms and conditions of the Creative Commons license CC BY-NC-ND.

Contents

About the Editors vii

Maria Teresa Cidade and João Miguel Nóbrega
Editorial for Special Issue "Advances in Experimental and Computational Rheology, Volume II"
Reprinted from: *Fluids* **2020**, *5*, 163, doi:10.3390/fluids5040163 1

Ruben Ibañez, Fanny Casteran, Clara Argerich, Chady Ghnatios, Nicolas Hascoet, Amine Ammar, Philippe Cassagnau and Francisco Chinesta
On the Data-Driven Modeling of Reactive Extrusion
Reprinted from: *Fluids* **2020**, *5*, 94, doi:10.3390/fluids5020094 3

J. Esteban López-Aguilar and Hamid R. Tamaddon-Jahromi
Computational Predictions for Boger Fluids and Circular Contraction Flow under Various Aspect Ratios
Reprinted from: *Fluids* **2020**, *5*, 85, doi:10.3390/fluids5020085 27

João Pedro, Bruno Ramôa, João Miguel Nóbrega and Célio Fernandes
Verification and Validation of *openInjMoldSim*, an Open-Source Solver to Model the Filling Stage of Thermoplastic Injection Molding
Reprinted from: *Fluids* **2020**, *5*, 84, doi:10.3390/fluids5020084 49

Raquel Portela, Filipe Valcovo, Pedro L. Almeida, Rita G. Sobral and Catarina R. Leal
Antibiotic Activity Screened by the Rheology of *S. aureus* Cultures
Reprinted from: *Fluids* **2020**, *5*, 76, doi:10.3390/fluids5020076 73

Cassio M. Oishi, Fernando P. Martins and Roney L. Thompson
Gravitational Effects in the Collision of Elasto-Viscoplastic Drops on a Vertical Plane
Reprinted from: *Fluids* **2020**, *5*, 61, doi:10.3390/fluids5020061 83

Luis G. Baltazar, Fernando M.A. Henriques and Maria Teresa Cidade
Effects of Polypropylene Fibers and Measurement Methods on the Yield Stress of Grouts for the Consolidation of Heritage Masonry Walls
Reprinted from: *Fluids* **2020**, *5*, 53, doi:10.3390/fluids5020053 101

Christine Macedo, Maria Cristiana Nunes, Isabel Sousa and Anabela Raymundo
Rheology Methods as a Tool to Study the Impact of Whey Powder on the Dough and Breadmaking Performance of Wheat Flour
Reprinted from: *Fluids* **2020**, *5*, 50, doi:10.3390/fluids5020050 115

Regina Miriam Parlato, Eliana R. Russo, Jörg Läuger, Salvatore Costanzo, Veronica Vanzanella and Nino Grizzuti
On the Use of the Coaxial Cylinders Equivalence for the Measurement of Viscosity in Complex Non-Viscometric, Rotational Geometries
Reprinted from: *Fluids* **2020**, *5*, 43, doi:10.3390/fluids5020043 129

Diana Alatalo and Fatemeh Hassanipour
An Experimental Study on Human Milk Rheology: Behavior Changes from External Factors
Reprinted from: *Fluids* **2020**, *5*, 42, doi:10.3390/fluids5020042 141

Yago Chamoun F. Soares, Elyff Cargnin, Mônica Feijó Naccache and Ricardo Jorge E. Andrade
Influence of Oxidation Degree of Graphene Oxide on the Shear Rheology of Poly(ethylene glycol) Suspensions
Reprinted from: *Fluids* **2020**, *5*, 41, doi:10.3390/fluids5020041 . . . . **163**

About the Editors

Maria Teresa Cidade is a member of the Polymeric and Mesomorphic Materials Group of the Faculty of Sciences and Technology of the New University of Lisbon (FCT NOVA), Portugal. She graduated in Chemical Engineering (IST/Technical University of Lisbon, 1983) and obtained a Ph.D. degree from FCT NOVA (1994). In 2006, she was appointed with the Habilitation in Polymer Engineering. Currently, she is an Assistant Professor with Habilitation at the Materials Science Department (DCM) of FCT NOVA. She is the Coordinator of the Polymeric and Mesomorphic Materials Group of DCM, Coordinator of the Rheology Sub-Group of the Soft and Bifunctional Materials Group of the Materials Research Centre (Cenimat) of DCM, Coordinator of the Doctoral Program in Materials Science and Engineering, and Coordinator of FCT NOVA in the Doctoral Program in Advanced Materials and Processing (a Doctoral Program in Association with six other Portuguese Universities: Lisbon, Coimbra, Beira Interior, Aveiro, Porto, and Minho, supported by the Portuguese Foundation for Science and Technology). She is the President of the Portuguese Society of Rheology, Associate Editor of *Physica Scripta* (IOP), and a member of the Editorial Board of *Fluids* (MDPI). Her main scientific interests include the rheology (including electrorheology and rheo-optics) of complex systems (polymers and polymeric base systems, liquid crystals, nanocomposites, biomaterials, building materials, etc.), the mechanical characterization of polymers and polymer composites, and polymer processing. During her career, she has supervised more than 30 researchers, coauthored four book chapters and 79 papers in international refereed journals, lodged 1 patent, and presented more than 100 communications in conferences.

João Miguel Nóbrega (Associate Professor) works at the Polymer Engineering Department of the University of Minho and is a member of the Institute for Polymers and Composites. In 2004, he received his Ph.D. degree from the University of Minho in Polymer Science and Engineering. He is the Editor of the OpenFOAM® Journal and OpenFOAM® Wiki, a founder member of the Iberian OpenFOAM® Technology Users, and the lead faculty of the Digital Transformation in Manufacturing area from the MIT Portugal Program. His research activities encompass three overlapping areas: product development, polymer processing, and material rheology. For this purpose, he has been developing computational rheology tools to model the flow of complex fluids in various polymer processing techniques. Regarding the product development area, he has been involved in the design and manufacture of polymeric products across several fields, comprising applications for health, textiles, sensoring/monitoring, construction, and mobility. In 2014, he joined the OpenFOAM® Extend community, and has focused, since then, on the main numerical developments in this open-source computational library. In 2016, he was the chair of the 11th Workshop OpenFOAM, which took place in Guimarães, Portugal. During his career, he was involved in the supervision of more than 50 researchers, working both in fundamental and applied research projects; he has coedited 4 books, and published more than 80 papers in international refereed journals and 22 book chapters, lodged 9 patents (3 international), and presented approximately 200 communications in conferences.

Editorial

Editorial for Special Issue "Advances in Experimental and Computational Rheology, Volume II"

Maria Teresa Cidade [1],* and João Miguel Nóbrega [2],*

1. Departamento de Ciência dos Materiais and Cenimat/I3N, Faculdade de Ciências e Tecnologia, Universidade Nova de Lisboa, 2829-516 Caparica, Portugal
2. Institute for Polymers and Composites, University of Minho, Campus de Azurém, 4800-058 Guimarães, Portugal
* Correspondence: mtc@fct.unl.pt (M.T.C.); mnobrega@dep.uminho.pt (J.M.N.)

Received: 21 September 2020; Accepted: 24 September 2020; Published: 25 September 2020

Rheology, defined as the science of the deformation and flow of matter, is a multidisciplinary scientific field, covering both fundamental and applied approaches. The study of rheology includes both experimental and computational methods, which are not mutually exclusive. Its practical relevance embraces many daily life processes, like preparing mayonnaise, spreading an ointment, or shampooing, and industrial processes, like polymer processing and oil extraction, among several others. Practical applications also include formulation and product development.

Following a successful first volume, the Special Issue entitled "Advances in Experimental and Computational Rheology", the editorial team decided to launch a second volume.

The Special Issue "Advances in Experimental and Computational Rheology, Volume II" comprises 10 papers covering some of the latest advances in the fields of experimental and computational rheology, applied to a diverse class of materials and processes, which can be grouped into three main topics: rheology [1–5], rheometry and processing [6,7], and theoretical modeling [8–10].

The characterization of rheological behavior is the main topic of five contributions, covering the following materials/systems: S-aureus cultures (Portela et al. [1]), in which antibiotic activity was screened by rheometry; natural hydraulic lime grouts filled with polypropylene fibers (Baltazar et al. [2]), with a particular focus on the effect of the measurement methods on the obtained yield stress; wheat flour dough (Macedo et al. [3]), where rheology was used as a tool to study the impact of whey powder addition on the dough and breadmaking performance; human milk (Alatalo et al. [4]), covering the influence of external factors on its characteristics; and graphene oxide/poly(ethylene glycol) suspensions (Soares et al. [5]), where the authors studied the influence of the oxidation degree of graphene oxide on the suspensions' shear rheology.

Two of the Special Issue papers are dedicated to rheometry and processing. Ibañez et al. [6] analyzed the ability of different machine learning techniques, able to operate under a low data limit, to create a model linking material and process parameters with the properties and performances of parts obtained by reactive polymer extrusion. Parlato et al. [7] applied the so-called Couette analogy concept, in order to achieve a reduction in the complex, non-viscometric rotational geometry to a virtual concentric cylinder analogue, allowing for the determination of the flow curve of non-Newtonian fluids in complex geometries.

Theoretical modeling is the main topic of the remaining three works. The work of Lopéz Aguilar et al. [8] put forward a modeling framework that was experimentally validated, with a focus on the circular abrupt contraction flow of two highly elastic constant shear viscosity Boger fluids, with various contraction ratio geometries. Pedro et al. [9] numerically studied the filling stage of thermoplastic injection molding with a solver implemented in the open-source computational library OpenFOAM® and compared the new solver performance and accuracy with a proprietary code. In the work of OIshi et al. [10], the authors studied the gravitational effects of elasto-viscoplastic drops colliding on vertical planes and proposed a classification for the observed behaviors.

Finally, it is very important to recognize and acknowledge the effort put forth by the large number of anonymous reviewers, which was essential to assuring the high quality of all the contributions of this Special Issue.

Conflicts of Interest: The authors declare no conflict of interest.

References

1. Portela, R.; Valcovo, F.; Almeida, P.L.; Sobral, R.G.; Leal, C.R. Antibiotic Activity Screened by the Rheology of S. aureus Cultures. *Fluids* **2020**, *5*, 76. [CrossRef]
2. Baltazar, L.G.; Henriques, F.M.A.; Cidade, M.T. Effects of Polypropylene Fibers and Measurement Methods on the Yield Stress of Grouts for the Consolidation of Heritage Masonry Walls. *Fluids* **2020**, *5*, 53. [CrossRef]
3. Macedo, C.; Nunes, M.C.; Sousa, I.; Raymundo, A. Rheology Methods as a Tool to Study the Impact of Whey Powder on the Dough and Breadmaking Performance of Wheat Flour. *Fluids* **2020**, *5*, 50. [CrossRef]
4. Alatalo, D.; Hassanipour, F. An Experimental Study on Human Milk Rheology: Behavior Changes from External Factors. *Fluids* **2020**, *5*, 42. [CrossRef]
5. Soares, Y.C.F.; Cargnin, E.; Naccache, M.F.; Andrade, R.J.E. Andrade. Influence of Oxidation Degree of Graphene Oxide on the Shear Rheology of Poly(ethylene glycol) Suspensions. *Fluids* **2020**, *5*, 41. [CrossRef]
6. Ibañez, R.; Casteran, F.; Argerich, C.; Ghnatios, C.; Hascoet, N.; Ammar, A.; Cassagnau, P.; Chinesta, F. On the Data-Driven Modeling of Reactive Extrusion. *Fluids* **2020**, *5*, 94. [CrossRef]
7. Parlato, R.M.; Russo, E.R.; Läuger, J.; Costanzo, S.; Vanzanella, V.; Grizzuti, N. On the Use of the Coaxial Cylinders Equivalence for the Measurement of Viscosity in Complex Non-Viscometric, Rotational Geometries. *Fluids* **2020**, *5*, 43. [CrossRef]
8. López-Aguilar, J.E.; Tamaddon-Jahromi, H.R. Computational Predictions for Boger Fluids and Circular Contraction Flow under Various Aspect Ratios. *Fluids* **2020**, *5*, 85. [CrossRef]
9. Pedro, J.; Ramôa, B.; Nóbrega, J.M.; Fernandes, C. Verification and Validation of openInjMoldSim, an Open-Source Solver to Model the Filling Stage of Thermoplastic Injection Molding. *Fluids* **2020**, *5*, 84. [CrossRef]
10. Oishi, C.M.; Martins, F.P.; Thompson, R.L. Gravitational Effects in the Collision of Elasto-Viscoplastic Drops on a Vertical Plane. *Fluids* **2020**, *5*, 61. [CrossRef]

© 2020 by the authors. Licensee MDPI, Basel, Switzerland. This article is an open access article distributed under the terms and conditions of the Creative Commons Attribution (CC BY) license (http://creativecommons.org/licenses/by/4.0/).

Article

On the Data-Driven Modeling of Reactive Extrusion

Ruben Ibañez [1], Fanny Casteran [2], Clara Argerich [1], Chady Ghnatios [3], Nicolas Hascoet [1], Amine Ammar [4], Philippe Cassagnau [2] and Francisco Chinesta [1,*]

- [1] PIMM, Arts et Métiers Institute of Technology, 151 Boulevard de l'Hôpital, 75013 Paris, France; Ruben.IBANEZ-PINILLO@ensam.eu (R.I.); clara.argerich_martin@ensam.eu (C.A.); nicolas.hascoet@ensam.eu (N.H.)
- [2] Univ-Lyon, Université Lyon 1, Ingénierie des Matériaux Polymères, CNRS, UMR5223, 15 Boulevard André Latarjet, 69622 Villeurbanne, France; fanny.casteran65@gmail.com (F.C.); philippe.cassagnau@univ-lyon1.fr (P.C.)
- [3] Notre Dame University-Louaize, P.O. Box 72, Zouk Mikael, Zouk Mosbeh, Lebanon; cghnatios@ndu.edu.lb
- [4] LAMPA, Arts et Métiers Institute of Technology, 2 Boulevard du Ronceray, 49035 Angers, France; Amine.AMMAR@ensam.eu
- * Correspondence: Francisco.Chinesta@ensam.eu

Received: 2 June 2020; Accepted: 10 June 2020; Published: 15 June 2020

Abstract: This paper analyzes the ability of different machine learning techniques, able to operate in the low-data limit, for constructing the model linking material and process parameters with the properties and performances of parts obtained by reactive polymer extrusion. The use of data-driven approaches is justified by the absence of reliable modeling and simulation approaches able to predict induced properties in those complex processes. The experimental part of this work is based on the in situ synthesis of a thermoset (TS) phase during the mixing step with a thermoplastic polypropylene (PP) phase in a twin-screw extruder. Three reactive epoxy/amine systems have been considered and anhydride maleic grafted polypropylene (PP-g-MA) has been used as compatibilizer. The final objective is to define the appropriate processing conditions in terms of improving the mechanical properties of these new PP materials by reactive extrusion.

Keywords: reactive extrusion; data-driven; machine learning; artificial engineering; polymer processing; digital twin

---

1. Introduction

Initially, the industry adopted virtual twins in the form of simulation tools that represented the physics of materials, processes, structures, and systems from physics-based models. These computational tools transformed engineering science and technology to offer optimized design tools and became essential in almost all industries at the end of the 20th century.

Despite of the revolution that Simulation Based Engineering—SBE—experienced, some domains resisted to fully assimilate simulation in their practices for different reasons:

- Computational issues related to the treatment of too complex material models involved in too complex processes, needing a numerical resolution difficult to attain. Some examples in polymer processing concern reactive extrusion or foaming, among many others.
- Modeling issues when addressing materials with poorly known rheologies, as usually encountered in multi-phasic reactive flows where multiple reactions occur.
- The extremely multi-parametric space defined by both the material and the process, where the processed material properties and performances strongly depend on several parameters related, for example in the case of reactive extrusion, to the nature of the reactants or the processing parameters, like the flow rate and viscosity, the processing temperature, etc.

In these circumstances, the use of data and the construction of the associated models relating the material and processing parameters to some quantities of interest—QoI, by using advanced artificial intelligence techniques, seems an appealing procedure for improving predictions, enhancing optimization procedures and enabling real-time decision making [1].

1.1. Data-Driven Modeling

Engineered artificial intelligence—EAI—concerns different data-science functionalities enabling: (i) multidimensional data visualization; (ii) data classification; (iii) modeling the input/output relationship enabling quantitative predictions; (iv) extracting knowledge from data; (v) explaining for certifying, and (vi) creating dynamic data-driven applications systems.

The present work aims at creating a model able to relate material and processing parameters to the processed material properties and performances. For this reason, in that follows, we will focus on the description and use of different strategies for accomplishing that purpose.

In the past, science was based on the extraction of models, these being simply the causal relation linking causes (inputs) and responses (outputs). This (intelligent) extraction or discovery was performed by smart (and trained) human minds from the data provided by the direct observation of the reality or from engineered experimental tests. Then, with the discovered, derived, or postulated model, predictions were performed, leading to the validation or rejection of these models. Thus, physics-based models, often in the form of partial differential equations, were manipulated by using numerical techniques, with the help of powerful computers.

However, sometimes models are not available, or they are not accurate enough. In that case, the most natural route consists of extracting the model from the available data (a number of inputs and their associated outputs). When data are abundant and the time of response is not a constraint, deep-learning could constitute the best alternative. However, some industrial applications are subjected to: (i) scarce data and (ii) necessity of learning on-the-fly under stringent real-time constraints.

Some models, as those encountered in mechanics, are subjected to thermodynamic consistency restrictions. They impose energy conservation and entropy production. In our former works [2–5], we proved that such a route constitutes a very valuable framework for deriving robust models able to assimilate available data while fulfilling first principles. However, some models cannot be cast into a physical framework because they involve heterogeneous data, sometimes discrete and even categorical. Imagine for awhile a product performance that depends on four factors: (i) temperature of the oven that produces the part; (ii) process time; (iii) commercial name of the involved material, and (iv) given-name of the employee that processed it. It seems evident that processes, whose representing data-points (implying four dimensions in this particular example) are close to each other, do not imply having similar performances. In that case, prior to employing techniques performing in vector spaces, in general based on metrics and distances, data must be mapped into that vector space. For this purpose, we proposed recently the so-called *Code2Vect* [6] revisited later.

Nonlinear regressions, relating a given output with the set of input parameters, are subjected to a major issue: complexity. In other words, the number of terms of a usual polynomial approximation depends on the number of parameters and the approximation degree. Thus, D parameters and a degree Q imply the order of D power Q terms and, consequently, the same amount of data are needed to define it. In our recent works, we proposed the so-called sparse Proper Generalized Decomposition, sPGD [7] able to circumvent the just referred issue. It is based on the use of separated representations with adaptive degree of approximation, defined on unstructured data settings and sparse sensing to extract the most compact representations.

Assuming that the model is expressible as a matrix relating arrays of inputs and outputs (as standard Dynamic Mode Decomposition—DMD—performs with dynamical systems [8,9]) both expressible in a low-dimensional space (assumption at the heart of all model order reduction techniques), the rank of the matrix (discrete description of the model) is assumed to be low [10].

1.2. Reactive Polymers Processing

Reactive extrusion is considered to be an effective tool of continuous polymerization of monomers, chemical modification of polymers and reactive compatibilisation of polymer blends. In particular, co-rotating and contra-rotating twin-screw extruders have proven to be a relevant technical and economical solution for reactive processing of thermoplastic polymers. The literature dedicated to reactive extrusion shows that a very broad spectrum of chemical reactions and polymer systems has been studied [11–15].

The many advantages of using the extruder as a chemical reactor can be described as follows: (i) polymerization and/or chemical modifications can be carried out in bulk, in the absence of solvents, the process is fast and continuous (residence time of the order of a few minutes); (ii) if necessary, devolatilization is effective, leading to the rapid removal of residual monomers and/or reaction by-products; and (iii) the screw design is modular, allowing the implementation of complex formulations (fillers, plasticizers, etc.).

However, there are also some disadvantages in using an extruder as a chemical reactor such as: (i) the high viscosity of the molten polymers, which lead to self-heating and therefore to side reactions (thermal degradation for example); (ii) the short residence time which limits reactive extrusion to fast reactions; and (iii) the difficulty of the scale up to industrial pilot and plants.

In terms of modeling and simulation, various strategies [16] can be considered as it needs to deal with a large number of highly nonlinear and coupled phenomena. Actually, the strategy of modeling depends on the objectives in terms of process understanding, material development from machine design or process optimization, and control. For example, in the case of free radical grafting of polyolefins, a two-phase stochastic model to describe mass transport and kinetics based on reactive processing data was proposed in [17].

Regarding process optimization, a simple 1D simulation approach provides a global description of the process all along the screws, whereas 3D models allow a more or less accurate description of the flow field in the different full zones of the extruder. However, most of these simulations are based on simplified steady-state 1D models (e.g., Ludovicc© software [18]).

Actually, the main processing parameters such as residence time, temperature, and extent of the reaction are assumed homogeneously distributed in any axial cross section. The use of one-dimensional models allows significant reductions of the simulation effort (computing time savings). In any case, the flow model is coupled with reaction kinetics that impact the fluid rheology [19].

Thus, one-dimensional models are specially appropriate when addressing optimization or control in reactive extrusion. In particular, the model proposed in [20] predicts the transient and steady-state behaviors, i.e., pressure, monomer conversion, temperature, and residence time distribution in different operation conditions.

However, these simulations require several sub-models on establishing constitutive equations (viscosity, chemical kinetics, mass and temperature transfers). Actually, it takes time and the intuition and accumulated knowledge of experienced specialists. Furthermore, it is important to note that, despite the impressive effort spent by hundreds of researchers and thousands of published papers, no constitutive equation exists describing, for example, the behavior of complex polymer formulations such as reactive extrusion systems.

In summary, such a process is quite complex and would require a detailed study on the influence of the nature of polymers and chemical reactions (kinetics and rheology), processing conditions (temperature, screw speed, flow rate, screw profile). Nevertheless, a determinist answer to each of these parameters is out of consideration and actually we believe that the understanding of such a process is quite unrealistic from usual approaches.

1.3. Objectives of the Study

The present work aims at addressing a challenge in terms of industrial applications that is not necessarily based on improving the understanding of the process itself, but replacing the complex fluid and complex flow by an alternative modeling approach able to extract the link between the process outputs and inputs, key for transforming experience into knowledge.

A model of a complex process could be envisaged with two main objectives: (i) the one related to the online process control from the collected and assimilated data; (ii) the other concerned by the offline process optimization, trying to extract the optimal process parameters enabling the target properties and performances. Even if the modeling procedure addressed in this work could be used in both domains, the present work mainly focuses on the second one, the process modeling for its optimization; however, as soon as data could be collected in real-time, with the model available, process control could be attained without major difficulties.

There are many works in which each one uses a different data-driven modeling technique, diversity that makes it difficult to understand if there is an optimal technique for each model, or if most of them apply and perform similarly. Thus, this paper aims at comparing several techniques first, and then, using one of them that the authors recently proposed, and that performs in the multi-parametric setting, address some potential uses.

2. Modeling

In this section, we revisit some regression techniques that will be employed after for modeling reactive extrusion. For additional details and valuable references the interested reader can refer to Appendix A.

In many applications like chemical and process engineering or materials processing, product performances depend on a series of parameters related to both, the considered materials and the processing conditions. The number of involved parameters is noted by D and each parameter by x_i, $i = 1, \ldots, D$, all of them grouped in the array $\mathbf{x}$.

The process results in a product characterized by different properties or performances in number smaller or greater than D. In what follows, for the sake of simplicity and without loss of generality, we will assume that we are interested in a single scalar output noted by y.

From the engineering point of view, one is interested in discovering the functional relation between the quantity of interest—QoI—y and the involved parameters $x_1, \ldots, x_D \equiv \mathbf{x}$, mathematically, $y = y(\mathbf{x})$ because it offers a practical and useful way for optimizing the product by choosing the most adequate parameters $\mathbf{x}^{\text{opt}}$.

There are many techniques for constructing such a functional relation, currently known as regression, some of them sketched below, and detailed in Appendix A where several valuable references are given.

2.1. From Linear to Nonlinear Regression

The simplest choice consists in the linear relationship

$$y = \beta_0 + \beta_1 x_1 + \cdots + \beta_D x_D, \tag{1}$$

that if $D + 1$ data are available that is $D + 1$ couples $\{y_s, \mathbf{x}_s\}$, $s = 1, \ldots, D + 1$, then the previous equation can be written in the matrix form

$$\begin{pmatrix} y_1 \\ y_2 \\ \vdots \\ y_{D+1} \end{pmatrix} = \begin{pmatrix} 1 & x_{1,1} & x_{2,1} & \cdots & x_{D,1} \\ 1 & x_{1,2} & x_{2,2} & \cdots & x_{D,2} \\ \vdots & \vdots & \vdots & \ddots & \vdots \\ 1 & x_{1,D+1} & x_{2,D+1} & \cdots & x_{D,D+1} \end{pmatrix} \begin{pmatrix} \beta_0 \\ \beta_1 \\ \vdots \\ \beta_D \end{pmatrix}, \tag{2}$$

where $x_{i,s}$ denotes the value of parameter x_i at measurement s, with $i = 1, \ldots, D$ and $s = 1, \ldots, D+1$. The previous linear system can be expressed in a more compact matrix form as

$$\mathbf{y} = \mathbb{X}\boldsymbol{\beta}. \tag{3}$$

Thus, the regression coefficients $\beta_0, \ldots, \beta_D$ are computed by simple inversion of Equation (3)

$$\boldsymbol{\beta} = \mathbb{X}^{-1}\mathbf{y}, \tag{4}$$

from which the original regression form (1) can be rewritten as

$$y = \beta_0 + \mathbf{W}^T \mathbf{x}, \tag{5}$$

where $\mathbf{W}^T = (\beta_1 \cdots \beta_D)$.

When the number of measurements P becomes larger than the number of unknowns $\beta_0, \cdots, \beta_D$, i.e., $P > D + 1$, the problem can be solved in a least-squares sense.

However, sometimes linear regressions become too poor for describing nonlinear solutions and in that case one is tempted to extended the regression (1) by increasing the polynomial degree. Thus, the quadratic counterpart of Equation (1) reads

$$y = \beta_0 + \sum_{i=1}^{D} \sum_{j \geq i}^{D} \beta_{ij} x_i x_j, \tag{6}$$

where the number of unknown coefficients (β_0 & β_{ij}, $\forall i, j$) is $(D^2 - D)/2$ that roughly scales with D^2. When considering third degree approximations, the number of unknown coefficients scales with D^3 and so on.

Thus, higher degree approximations are limited to cases involving few parameters, and multi-parametric cases must use low degree approximations because usually the available data are limited due to the cost of experiences and time.

The so-called sparse-PGD [7] tries to encompass both wishes in multi-parametric settings: higher degree and few data. For that purpose, the regression reads

$$y = \sum_{i=1}^{N} \prod_{j=1}^{D} F_i^j(x_j), \tag{7}$$

where the different single-valued functions $F_i^j(x_j)$ are a priori unknown and are determined sequentially using an alternate directions fixed point algorithm. As at each step one looks for a single single-valued function, higher degree can be envisaged for expressing it into a richer (higher degree) approximation basis, while keeping reduced the number of available data-points (measurements).

2.2. Code2Vect

This technique deeply revisited in the Appendix A proposes mapping points $\mathbf{x}_s$, $s = 1, \ldots, P$, into another space $\boldsymbol{\xi}_s$, such that the distance between any pair of data-points $\boldsymbol{\xi}_i$ and $\boldsymbol{\xi}_j$ scales with the difference of their respective outputs, that is, on $|y_i - y_j|$.

Thus, using this condition for all the data-point pairs, the mapping $\mathbf{W}$ is obtained, enabling for any other input array $\mathbf{x}$ compute its image $\boldsymbol{\xi} = \mathbf{Wx}$. If $\boldsymbol{\xi}$ is very close to $\boldsymbol{\xi}_s$, one can expect that its output y becomes very close to y_s, i.e., $y \approx y_s$. In the most general case, an interpolation of the output is envisaged.

2.3. iDMD, Support Vector Regression, and Neural Networks

Inspired by dynamic model decomposition—DMD—[8,9] one could look for **W** minimizing the functional $\mathcal{F}(\mathbf{W})$ [10]

$$\mathcal{F}(\mathbf{W}) = \sum_{s=1}^{P} \left(y_s - \mathbf{W}^T\mathbf{x}_s\right)^2, \tag{8}$$

whose minimization results in the calculation of vector **W** that at its turn allows defining the regression $y = \mathbf{W}^T\mathbf{x}$. Appendix A and the references therein propose alternative formulations.

Neural Networks—NN—perform the same minimization and introduce specific treatments of the nonlinearities while addressing the multi-output by using a different number of hidden neuron layers [21].

Finally, Support Vector Regression—SVR—share some ideas with the so-called Support Vector Machine—SVM [22], the last widely used for supervised classification. In SVR, the regression reads

$$y = \beta_0 + \mathbf{W}^T\mathbf{x}, \tag{9}$$

and the flatness in enforced by minimizing the functional $\mathcal{G}(\mathbf{W})$

$$\mathcal{G}(\mathbf{W}) = \frac{1}{2}\mathbf{W}^T\mathbf{W}, \tag{10}$$

while enforcing as constraints a regularized form of

$$|y_s - \beta_0 - \mathbf{W}^T\mathbf{x}_s| \leq \epsilon, \quad s = 1, \ldots, P. \tag{11}$$

3. Experiments

The purpose of this project is the dispersion of a thermosetting (TS) polymer in a polyolefin matrix using reactive extrusion by in situ polymerisation of the thermoset (TS) phase from an expoxide resin and amine crosslinker. Here, Polypropylene (PP) has been chosen as the polyolefin matrix. A grafted PP maleic anhydride (PP-g-MA) has been used to ensure a good compatibility between the PP and the thermoset phases.

These studies were carried out as part of a project with TOTAL on the basis of a HUTCHINSON patent [23]. This patent describes the process for preparing a reinforced and reactive thermoplastic phase by dispersing an immiscible reactive reinforcing agent (e.g., an epoxy resin as precursor on the thermoset dispersed phase). This process is characterized by a high shear rate in the extruder combined with the in-situ grafting, branching, and/or crosslinking of the dispersed phase. These in situ reactions permit the crosslinking of the reinforcing agent as well as the compatibility of the blend with or without compatibilizer or crosslinker. The result of this process is a compound with a homogeneous reinforced phase with thin dispersion (<5 μm) leading to an improvement of the mechanical properties of the thermoplastic polymer. The experiments carried out in the framework of the present project are mainly based on some experiments described in the patent. However, new complementary experiments have been carried out to complete the study.

3.1. Materials

The main Polypropylene used as the matrix is the homopolymer polypropylene PPH3060 from TOTAL. Two other polypropylenes have been used to study the influence of the viscosity, and several impact copolymer polypropylenes have also been tested in order to combine a good impact resistance with the reinforcement brought by the thermoset phase. A PP-g-MA (PO1020 from Exxon) with around 1 wt% of maleic anhydride has been used as a compatibilizer between the polypropylene matrix and the thermoset phase. All the polypropylenes used are listed in Table 1 with their main characteristics.

Table 1. Nature, supplier, and melt flow index –MFI– (216 kg/230 °C/10 min) of PP polymers.

Name	Nature	Supplier	MFI (g)
PPH3060	Polypropylene homopolymer	TOTAL	1.8
PPH7060	Polypropylene homopolymer	TOTAL	12
PPH10060	Polypropylene homopolymer	TOTAL	35
PPC14642	Impact Copolymer Polypropylene	TOTAL	130
PPC10641	Impact Copolymer Polypropylene	TOTAL	44
PPC7810	Impact Copolymer Polypropylene	TOTAL	15
PPC7810C	Impact Copolymer Polypropylene	TOTAL	15
Exxelor PO1020	Polypropylene grafted 1 wt% of maleic anhydride	Exxon Mobil	430

Concerning the thermoset phase, three systems have been studied. As a common point, these three systems are based on epoxy resins that are DGEBA derivates with two epoxide groups, two different resins (DER 667 and DER 671 from DOW Chemicals have been used. The first two systems, named R1 and R2 here, are both constituted of an epoxy resin mixed with an amine at the stoichiometry. The first uses the DER 667 with a triamine (Jeffamine T403 from Huntsman) that is sterically hindered, whereas the second one uses the DER 671 with a cyclic diamine (Norbonanediamine from TCI Chemicals. Melamine has also been tested in one of the formulations. The third system, named R5 here, mixes the epoxy resin DER 671 with a phenolic hardener (DEH 84 from DOW Chemicals) that is a blend of three molecules: 70 wt% of an epoxy resin, a diol, and less than 1 wt% of a phenolic amine. These systems have been chosen in order to see the influence of the structure, molar mass, and chemical nature on the in-situ generation of the thermoset phase within our polyolefin matrix. Table 2 summarizes the systems studied.

Table 2. Composition at the stoichiometry of the systems studied.

Epoxy	Amine/Hardener	Abbreviation
DER 667	Jeffamine T403	R1
DER 671	DEH 84	R2
DER 671	Norbonane diamine	R5

The kinetics of these chemical systems have been studied from the variation of the complex shear modulus from a time sweep experiment with an ARES-G2 Rheometer (TA Instruments). The experiments have been performed for temperatures from 115 °C to 165 °C using a 25 mm plate-plate geometry, with a 1 mm gap, at the frequency ω = 10 rad/s and a constant strain of 1%. The kinetics have been performed on a stoichiometric premix of the reactants. The gel times of the systems have thus been identified as the crossover point between the loss and storage modulus. Note that the reaction is too fast to be performed at temperatures beyond T = 165 °C. Consequently, an extrapolation according to an Arrhenius law allowed us to determine the gel time of the systems at T = 200 °C (Barrel temperature of the extruder). The results give a gel time lower than 10 s for the three systems (t_{gel}(R1) = 4.5 s, t_{gel}(R2) = 10 s, and t_{gel}(R5) < 1 s), so we made the hypothesis that the reaction time is much lower than 1 min and thus that the reaction is totally completed at the die exit of the extruder. Moreover, a Dynamic Mechanical Analysis (DMA) showed that the main mechanical relaxation T_α associated with the T_g of the thermoset phase is close to 80 °C, which is the T_g observed for TS bulk systems.

The influence of the addition of silica on the final properties has been studied with two different silicas (Aerosil R974 and Aerosil 200).

3.2. Methods

3.2.1. Extrusion Processing

The formulations have been fulfilled in one single step with a co-rotating twin screw extruder (Leistritz ZSE18, L/D = 60, D = 18 mm), with the screw profile described in Figure 1.

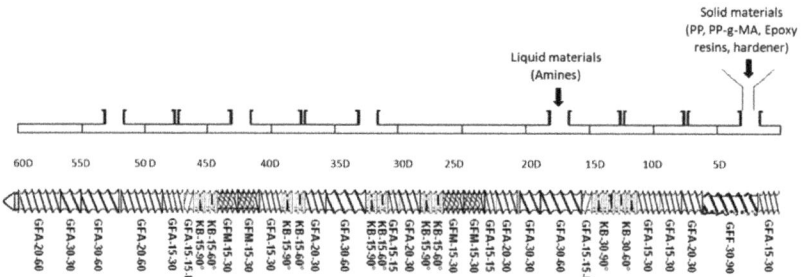

Figure 1. Diagram of the screw profile in the study.

Two different temperatures profiles have been used, one at 230 °C and the other one at 200 °C, both with lower temperatures for the first blocs to minimize clogging effects at the inlet. These temperature profiles are described in Figure 2.

230 °C Temperature profile											
N° Bloc	11	10	9	8	7	6	5	4	3	2	1	Alim
Setpoint Temperature	230 °C	230 °C	230 °C	230 °C	230 °C	230 °C	230 °C	230 °C	215 °C	190 °C	180 °C	110 °C

200 °C Temperature profile											
N° Bloc	11	10	9	8	7	6	5	4	3	2	1	Alim
Setpoint Temperature	200 °C	200 °C	200 °C	200 °C	200 °C	200 °C	200 °C	200 °C	200 °C	150 °C	150 °C	100 °C

Figure 2. Diagram of the temperature profile in the study.

Several screw rotation speeds and flow-rates have been used to study the influence of the process on the final materials (N = 300, 600, 450, 800 rpm; $\dot{w}$ = 3, 5, 6, 10 kg/h).

The solid materials were mixed and introduced at the entrance by a hopper for the pellets and with a powder feeder for the micronized powders. As for the liquid reagents, they were injected over the third bloc with an HPLC pump. The formulations are cooled by air at the exit of the extruder and then pelleted.

3.2.2. Characterization

Tensile-test pieces (5A) and impact-test pieces have been injected with a Babyplast injection press at 200 °C and 100 bar. Young modulus has been determined by a tensile test with a speed of 1 mm/min and Stress at yield, Elongation at break, and Stress at break have been measured with a tensile speed of 50 mm/min. Impact strength has been measured by Charpy tests on notched samples at room temperature.

4. Data-Driven Modeling: Comparing Different Machine Learning Techniques

As previously mentioned, a model that links the material and processing parameters with the processed material properties is of crucial interest. By doing that, two major opportunities could be envisaged: the first one concerns the possibility of inferring the processed material properties for any choice of manufacturing parameters; (second), for given target properties, one could infer the processing parameters enabling them.

In this particular case, process parameters are grouped in the six-entrees array $\mathbf{x}$:

$$\mathbf{x} = \begin{pmatrix} \text{Rotation Speed [rpm]} \\ \text{Exit Flow Rate [kg/h]} \\ \text{Temperature [°C]} \\ \text{TS dispersed Phase [\%]} \\ \text{PP} - \text{g} - \text{MA/TS Ratio [}-\text{]} \\ \text{Ratio TS [}-\text{]} \end{pmatrix} \quad (12)$$

whereas the processed material properties are grouped in the five-entrees array $\mathbf{y}$, containing the Young modulus, the yield stress, the stress at break, the strain at break, and the impact strength,

$$\mathbf{y} = \begin{pmatrix} E \, [MPa] \\ \sigma_y \, [MPa] \\ \sigma_b \, [MPa] \\ \epsilon_b \, [-] \\ Re_{amb} \, [\frac{kJ}{m^2}] \end{pmatrix}. \quad (13)$$

As previously discussed, our main aim is extracting (discovering) the regression relating inputs (material and processing parameters) $\mathbf{x}$ with the outputs (processed material properties) $\mathbf{y}$, the regression that can be written in the form

$$y_i = y_i(\mathbf{x}), \quad (14)$$

where $y_i()$ represents the linear or nonlinear regression associated with the i-output, or when proceeding in a compact form by creating the multi-valued regression relating the whole input and output data-pairs, as

$$\mathbf{y} = \mathbf{y}(\mathbf{x}). \quad (15)$$

The intrinsic material and processing complexity justify the nonexistence of valuable and reliable models based on physics, able to predict the material evolution and the process induced properties. For this reason, in the present work, the data-driven route is retained, from the use of regression techniques, as the ones previously summarized.

The available data come from experiments conducted, described in the previous section that consists of P pairs of arrays $\mathbf{x}_s$, and $\mathbf{y}_s$, $s = 1, \cdots, P$, that is:

$$[\mathbf{x}_s, \mathbf{y}_s], \quad s = 1, \ldots, P, \quad (16)$$

all them reported in Tables A1 and A2 included in the Appendix B (for the sake of completeness and for allowing researchers to test alternative regression procedures).

Table A1 groups the set of input parameters involved in the regression techniques. The hyper parameter *MaskIn* is a boolean mask indicating if the data are included in the training (*MaskIn*= 1) or it is excluded from the training to serve for quantifying the regressions performance (*MaskIn*= 0). On the other hand, Table A2 groups the responses, experimental measures, for each processing condition.

As indicated in the introduction section, one of the objectives of the present paper is analyzing if different machine learning techniques perform similarly, or their performances are significantly different. For this purpose, this section aims at comparing the techniques introduced in Section 2, whereas the next section will focus on the use of one of them.

In order to compare the performances of the different techniques, an error was introduced serving to compare the regressions prediction. In particular, we consider the most standard error, the Root

Mean Squared Error (RMSE). When applied to the different regression results, it offers a first indication on the prediction performances. Table 3 reports the errors associated with each regression when evaluating the output of interest that is the array **y** for a given input **x**, for all the data reported in Tables A1 and A2. Because the different outputs (the components of array **y**) present significant differences in their typical magnitudes, Table 4 reports the relative errors, computed from the ratios between the predicted and measured data difference to the measured data.

Table 3. RMSE error for the different regression techniques.

RMSE ERROR (%)	E (MPa)	σ_{yield} (MPa)	σ_{break} (MPa)	ϵ_{break}	Re_{amb} (kJ/m)
NN	369.81	7.50	6.92	119.16	2.44
C2V	289.22	5.91	5.81	126.83	1.89
SPGD	229.69	5.34	4.53	127.36	2.05
SVM	229.69	5.34	4.53	127.36	2.05
D. TREE	268.61	5.39	5.53	111.86	2.21
iDMD	252.72	5.41	5.53	114.83	1.89

Table 4. Relative error for different regression techniques.

REL ERROR (%)	E (MPa)	σ_{yield} (MPa)	σ_{break} (MPa)	ϵ_{break}	Re_{amb} (kJ/m)
NN	23.43	24.05	34.96	100.22	43.38
C2V	15.06	16.41	22.22	105.54	37.05
SPGD	18.32	18.97	29.38	106.67	33.73
SVM	14.55	17.12	22.90	107.12	36.58
D. TREE	17.02	17.30	27.96	94.08	39.38
iDMD	16.01	17.35	27.96	96.58	33.70

Sparse PGD—sPGD—employed second degree Chebyshev polynomials and performed a regression for each of the quantities of interest according to Equation (14). The use of low degree polynomials avoided overfitting, being a compromise for ensuring a reasonable predictability for data inside and outside the training data-set. From a computational point of view, 20 enrichment (N in Equation (A16)) were needed for defining the finite sum involved in the separated representation that constitutes the regression of each output of interest y_i.

Code2Vect addressed the low-data limit constraint by imposing a linear mapping between the representation (data) and target (metric) spaces, avoiding spurious oscillations when making predictions on the data outside the training set.

Considering the iDMD because of the reduced amount of available data, the simplest option consisting of a unique matrix relating the input–output data pairs in the training set (linear model) was considered, i.e., with respect to Equation (15), it was assumed $\mathbf{F}(\mathbf{x})) = \mathbb{F}\mathbf{x}$, and matrix $\mathbb{F}$ ensuring the linear mapping was obtained by following the rationale described in Section 2. The computed regression performs very well despite the fact of assuming a linear behavior.

The quite standard Neural Network we considered (among a very large variety of possible choices) presents a severe overfitting phenomenon in the low data-limit addressed here. This limitation is not intrinsic to NN, and could be alleviated by considering richer architectures and better optimizers, parameters, out of the scope of the present study.

The main conclusion of this section is the fact that similar results are obtained independently of the considered technique that seems quite good now from the point of view of engineering. Even if the errors seem quite high, it is important to note that: (i) the highest errors concern the variables exhibiting the largest dispersion in the measurements; (ii) the prediction errors are of the same order as that of the dispersion amplitudes; and (iii) we only considered 35 data-points from the 59 available for the training (regression construction) while the reported errors were calculated by using the whole available data (the 59 data-points). The next section proves that the prediction quality increases with the increase of the amount of points involved in the regression construction (training).

5. Data-Driven Process Modeling

In view of the reported results, it can be stressed that all the analyzed techniques show similar performances and work reasonably well in the low-data limit (where only 60% out of 59 data points composed the training data-set were used in the regressions).

As it can be noticed, some quantities of interest such as the Young's modulus and the stress at break are quite well predicted when compared with the others on which predictions were less performant. There is a strong correlation between this predictability capability and the experimental dispersion noticed when measuring these other quantities, like the strain at break. That dispersion represents without any doubt a limit in the predictability that must be addressed within a probabilistic framework. All the mechanical tests were performed on five samples from the same experiment process on the extruder. The final value is the average of these five tests. The confidence interval is estimated: 10% for the Young modulus and Yield stress, 20% on elongation and stress at break.

Extracting a model of a complex process could serve for real-time control purposes, but also, as it is the case in the present work, for understanding the main tendencies of each quantity of interest with respect to each process or material parameter (the last constituting the regression inputs), enabling process and material optimization.

In order to perform that sensibility analysis, we consider a given quantity of interest and evaluate its evolution with respect to each of the input parameters. When considering the dependence with respect to a particular input parameter, all the others are fixed to their mean values, even if any other choice is possible. Figure 3 shows the evolution of σ_b with respect to the six input parameters, using the lowest order sPGD modes to extract the main tendencies.

From these AI-based metamodels, one should be able to identify the process conditions and the concentration of the TS phase in order to enhance a certain mechanical property. Thus, in order to increase the stress at break, increasing the content of thermoset seems a good option, with all the other properties (Young modulus, stress at yield, strain at break and impact strength being almost insensible to that parameter). A more detailed analysis, involving multi-objective optimization (making use of the Pareto front) and its experimental validation, constitutes a work in progress, out of the scope of the present work.

To further analyze the accuracy of the methodology and the convergence behavior, in what follows, we consider one of the regression techniques previously described and employed, the sPGD, and perform a convergence analysis, by evaluating the evolution of the error with respect to the size of the training data-set.

The training-set was progressively enriched, starting from 30 data points, and then considering 35, 41, 47, and finally 53 (that almost correspond to 50%, 60%, 70%, 80%, and 90% of the available data-set). The error was calculated again by considering both the training and test data-sets. Table 5 reports the results on the elastic modulus prediction, and clearly proves, as expected, that the prediction accuracy increases with the size of the training-set, evolving from around 15% to finish slightly below 10%.

It is important to note that one could decrease even more the error when predicting the training-set, but overfitting issues will occur and the error will increase tremendously out of the training set compromising robustness. The errors here reported are a good compromise between accuracy in and out of the training-set.

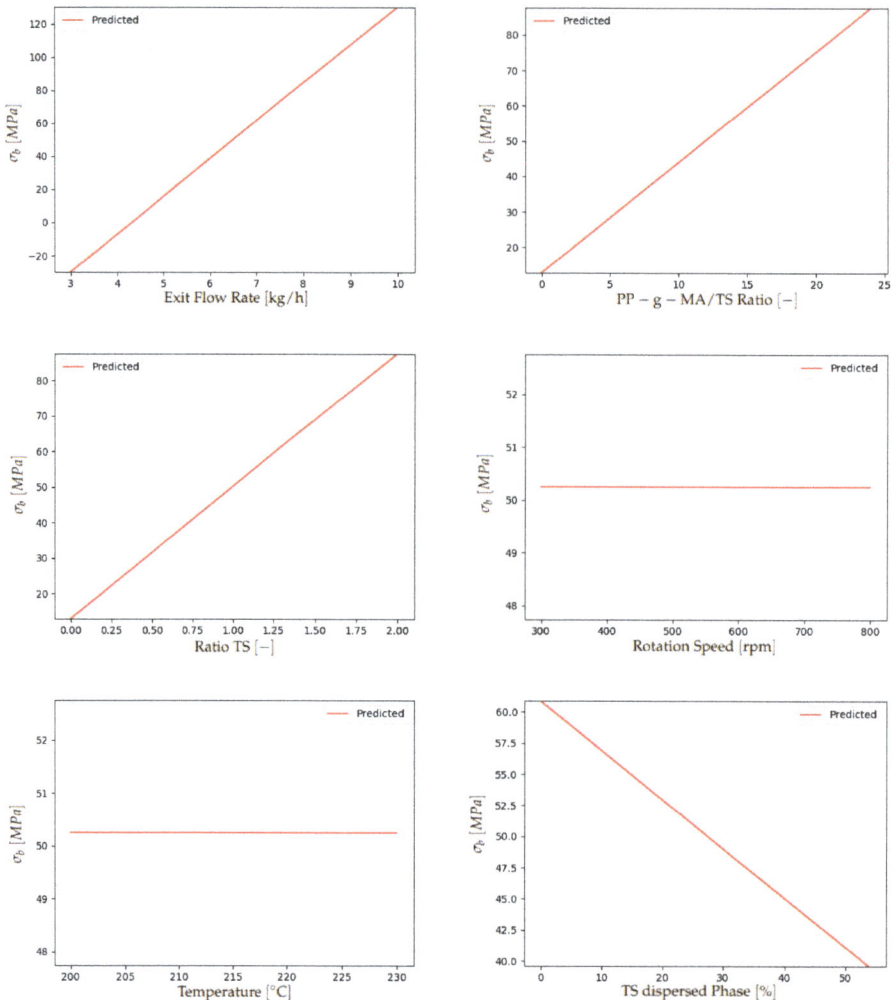

Figure 3. Evolution of the stress at break with: (**top-left**) Exit flow rate; (**top-right**) PP-g-MA/TS ratio; (**middle-left**) Ratio TS; (**middle-right**) Rotation speed; (**bottom-left**) Temperature; and (**bottom-right**) TS dispersed.

Table 5. Mean error for different numbers of points in the training-set.

Training-Set	Mean Error (%)
30	15.4
35	12.0
41	10.4
47	10.1
53	9.9

In order to facilitate the solution reproducibility, in what follows, we give the explicit form of the sPGD regression. As previously discussed, the sPGD makes use of a separated representation of the parametric solution, whose expression reads for a generic quantity of interest $u(\mathbf{x})$

$$u(\mathbf{x}) = \sum_{i=1}^{N} \prod_{j=1}^{D} F_i^j(x_j). \tag{17}$$

More explicitly, each univariate funcion $F_i^j(x_j)$ is approximated using an approximation basis,

$$u(\mathbf{x}) = \sum_{i=1}^{N} \prod_{j=1}^{D} \sum_{k=1}^{Q} T_k^j(x_j) \alpha_{ki}^j. \tag{18}$$

When approximating the elastic modulus, whose results were reported in Table 5, we considered six parameters, i.e., $D = 6$, a polynomial Chebyshev basis consisting of the functions $T_k^j(x_j)$ (needing for a pre-mapping of the parameter intervals into the reference one $[-1, 1]$ where the Chebyshev polynomials are defined). The number of modes (terms involved by the finite sum separated representation) and number of interpolation functions per dimension were set to $N = 10$ and $Q = 3$. The coefficients related to Equation (18) when applied to the elastic modulus approximation are reported in Appendix C.

An important limitation, inherent to machine learning strategies, is the fact that most likely other factors instead of the ones considered as inputs could be determinant for expressing the selected outputs. This point constitutes a work in progress.

6. Conclusions

We showed in this paper that different machine learning techniques are relevant in the low-data limit, for constructing the model that links material properties and process parameters in reactive polymer processing. Actually, these techniques are undeniably effective in complex processes such as reactive extrusion. More precisely, this work was based on the in situ synthesis of a thermoset phase during its mixing/dispersion with a thermoplastic polymer phase, which is certainly one of the most complex cases in the processing of polymers.

We proved that a variety of procedures can be used for performing the data-driven modeling, whose accuracy increases with the size of the training-set. Then, the constructed regression can be used for predicting the different quantities of interest, for evaluating their sensitivity to the parameters, crucial for offline process optimization, and also for real-time process monitoring and control.

Author Contributions: Conceptualization, P.C. and F.C. (Francisco Chinesta); methodology, R.I. and F.C. (Fanny Casteran); software, R.I., C.A., C.G., N.H., and A.A.; validation, F.C. (Fanny Casteran). All authors have read and agreed to the published version of the manuscript.

Funding: This research received no external funding.

Acknowledgments: Authors acknowledge ESI Group by its research chair at Arts et Metiers Institute of Technology, the French ANR through the DataBEST project as well as Nicolas Garois (Hutchinson company) for initiating the experimental part of this project with the support of the Total Research Department (La Defense, Paris). Mathilde Aucler, Aurelie Vanhille and Luisa Barroso-Gago are thanked for their help with the extruder and mechanical experiments.

Conflicts of Interest: The authors declare no conflict of interest.

Appendix A. Machine Learning Techniques

This appendix revisits, with some technical detail, the different machine learning techniques employed in the analyzes addressed in the present work.

Appendix A.1. Support Vector Regression

Finally, Support Vector Regression—SVR—shares some ideas of the so-called Support Vector Machine—SVM—widely used in supervised classification. In SVR, the regression reads

$$y = \beta_0 + \mathbf{W}^T\mathbf{x}, \tag{A1}$$

and the flatness in enforced by minimizing the functional $\mathcal{G}(\mathbf{W})$

$$\mathcal{G}(\mathbf{W}) = \frac{1}{2}\mathbf{W}^T\mathbf{W}, \tag{A2}$$

while enforcing as constraints a regularized form of

$$|y_s - \beta_0 - \mathbf{W}^T\mathbf{x}_s| \leq \epsilon, \quad s = 1, \ldots, P, \tag{A3}$$

in particular

$$\mathcal{G}(\mathbf{W}) = \frac{1}{2}\mathbf{W}^T\mathbf{W} + C\sum_{s=1}^{P}\xi_s\xi_s^*, \tag{A4}$$

with $\xi_s \geq 0$ and $\xi_s^* \geq 0$, and

$$y_s - (\mathbf{W}^T\mathbf{x}_s + \beta_0) \leq \epsilon + \xi_s, \tag{A5}$$

and

$$(\mathbf{W}^T\mathbf{x}_s + \beta_0) - y_s \leq \epsilon + \xi_s^*, \tag{A6}$$

with many other more sophisticated alternatives to extend to the nonlinear case.

Appendix A.2. Code-to-Vector—Code2Vect

Code2Vect maps data, eventually heterogenous, discrete, categorial, etc. into a vector space equipped of a Euclidean metric allowing computing distances, and in which points with similar output y remain close one to one another as sketched in Figure A1.

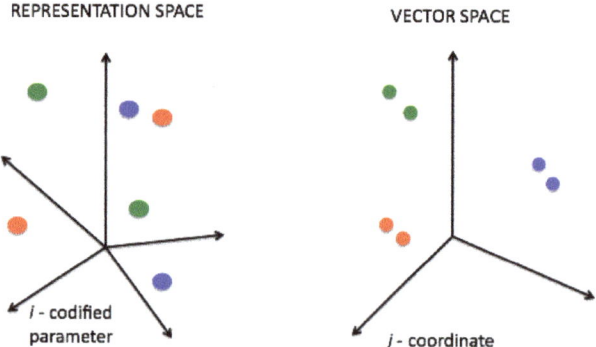

Figure A1. Input space (**left**) and target vector space (**right**) [6].

We assume that points in the origin space (space of representation) consist of P arrays composed on D entries, noted by $\mathbf{x}_i$. Their images in the vector space are noted by $\boldsymbol{\xi}_i \in \mathbb{R}^d$, with $d \ll D$. The mapping is described by the $d \times D$ matrix $\mathbf{W}$,

$$\boldsymbol{\xi} = \mathbf{W}\mathbf{x}, \tag{A7}$$

where both the components of $\mathbf{W}$ and the images $\boldsymbol{\xi}_i \in \mathbb{R}^d$, $i = 1, \cdots, P$, must be calculated. Each point $\boldsymbol{\xi}_i$ keeps the label (value of the output of interest) associated with its origin point $\mathbf{x}_i$, denoted by y_i.

We would like to place points $\boldsymbol{\xi}_i$, such that the Euclidian distance with each other point $\boldsymbol{\xi}_j$ scales with their output difference, i.e.,

$$(\mathbf{W}(\mathbf{x}_i - \mathbf{x}_j)) \cdot (\mathbf{W}(\mathbf{x}_i - \mathbf{x}_j)) = \|\boldsymbol{\xi}_i - \boldsymbol{\xi}_j\|^2 = |y_i - y_j|, \tag{A8}$$

where the coordinates of one of the points can be arbitrarily chosen. Thus, there are $\frac{P^2}{2} - P$ relations to determine the $d \times D + P \times d$ unknowns.

Linear mappings are limited and do not allow with proceeding in nonlinear settings. Thus, a better choice consists of the nonlinear mapping $\mathbf{W}(\mathbf{x})$ [6].

Appendix A.3. Incremental DMD

We reformulate the identification problem in a general multipurpose matrix form

$$\mathbb{W}\mathbf{x} = \mathbf{y}, \tag{A9}$$

where $\mathbf{x}$ and $\mathbf{y}$ represent the input and output vectors, involving variables of different nature, both them accessible from measurements. In what follows, both are assumed D-component arrays.

If we assume both evolving in low-dimensional sub-spaces of dimension d, with $d \ll D$, the rank of $\mathbb{K}$, the so-called model, is expected reducing to d. The construction of such reduced model was reported in [10], and from the two procedures that were proposed, in what follows, we summarize one of them, the so-called Progressive Greedy Construction.

In this case, we proceed progressively. We consider the first available datum, the pair $(\mathbf{x}_1, \mathbf{y}_1)$. Thus, the first, one-rank, reduced model reads

$$\mathbb{W}_1 = \frac{\mathbf{y}_1 \mathbf{y}_1^T}{\mathbf{y}_1^T \mathbf{x}_1}, \tag{A10}$$

ensuring $\mathbb{W}_1 \mathbf{x}_1 = \mathbf{y}_1$.

Suppose now that a second datum arrives $(\mathbf{x}_2, \mathbf{y}_2)$, from which we can also compute its associated rank-one approximation, and so on, for any new datum $(\mathbf{x}_i, \mathbf{y}_i)$:

$$\mathbb{W}_i = \frac{\mathbf{y}_i \mathbf{y}_i^T}{\mathbf{y}_i^T \mathbf{x}_i}, \quad i = 1, \cdots, P. \tag{A11}$$

For any other $\mathbf{x}$, the model could be interpolated from the just defined rank-one models, $\mathbb{W}_i$, $i = 1, ..., P$, according to

$$\mathbb{W}|_\mathbf{x} \approx \sum_{i=1}^{P} \mathbb{W}_i \mathcal{I}_i(\mathbf{x}), \tag{A12}$$

with $\mathcal{I}_i(\mathbf{x})$ the interpolation functions operating in the space of the data $\mathbf{x}$, functions that in general decrease with the distance between $\mathbf{x}$ and $\mathbf{x}_i$ (e.g., polynomials, radial basis, ...) able to proceed in multidimensional settings.

Appendix A.4. From Polynomial Regression to Sparse PGD-Based Regression

In the regression setting, one could consider a polynomial dependence of the QoI, y, on the parameters $x_1, \cdots, x_D$. The simplest choice, linear regression, reads

$$y(\mathbf{x}) = \beta_0 + \beta_1 x_1 + \cdots + \beta_D x_D, \tag{A13}$$

where the $D+1$ coefficients β_k can be computed from the available data. If $1+D=P$ data are available, $y_j, j = 1, \ldots, 1+D$, coefficients β_k can be calculated.

Linear regression requires the same amount of data as the number of involved parameters; however, it is usually unable to address nonlinearities.

Nonlinear regressions can be envisaged when the number of parameters remains reduced, due to the fact that the number of terms roughly scales with D to the power of the considered approximation degree.

In this section, we propose a technique able to ensure rich approximations while keeping the sampling richness quite reduced, the so-called multi-local sparse nonlinear PGD-based regression—sPGD. The last reads

$$\int_\Omega w(\mathbf{x})\,(y(\mathbf{x})-y)\,d\mathbf{x} = 0, \quad \forall w(\mathbf{x}) \tag{A14}$$

where Ω is the domain in the parametric space in which the approximation is searched, i.e., $\mathbf{x} \in \Omega$, and $w(\mathbf{x})$ represents the test function whose arbitrariness serves to enforce that the regression $y(\mathbf{x})$ approximates the available data y, and

$$y = \sum_{j=1}^{P} y\delta(\mathbf{x}_j), \tag{A15}$$

with δ the Dirac mass, to express that data only available at locations $\mathbf{x}_j$ in the parametric space.

Following the Proper Generalized Decomposition (PGD) rationale, the next step is to express the approximated function $y(\mathbf{x})$ in the separated form

$$y(\mathbf{x}) \approx \sum_{i=1}^{N} F_i^1(x_1) F_i^2(x_2) \cdots, \tag{A16}$$

constructed by using the standard rank-one update [7] that leads to the calculation of the different functions $F_i^j(x_j)$ involved in the separated form (A16).

Appendix A.5. A Simple Neural Network

Deep-learning is mostly based on the use of neural networks, networks composed of components that emulate the neuron functioning that from some incoming data generates an output that, within more complex and large networks, can become the input of other neurons in other layer.

We consider the schema in Figure A2 that illustrates a neuron receiving two input data x_1 and x_2 to produce the output Y. The simplest functioning consists of collecting both data, multiplying each by a weight, W_1 and W_2 and generate the output by adding both contributions according to

$$y = W_1 x_1 + W_2 x_2, \tag{A17}$$

that in the more general case can be written as

$$y = \sum_{i=1}^{D} W_i x_i \tag{A18}$$

or

$$y = \mathbf{W}^T \mathbf{x}. \tag{A19}$$

The main issue is precisely the determination of vector **W**. If an input–output couple is available $(\mathbf{x}_1, y_1)$, with the input normalized, i.e., $\|\mathbf{x}_1\| = 1$, then the best choice for the searched vector consists of $\mathbf{W} = y_1 \mathbf{x}_1$ that ensures recovering the known output, i.e.,

$$\mathbf{W}^T \mathbf{x}_1 = \left(y_1 \mathbf{x}_1^T \right) \mathbf{x}_1 = y_1. \tag{A20}$$

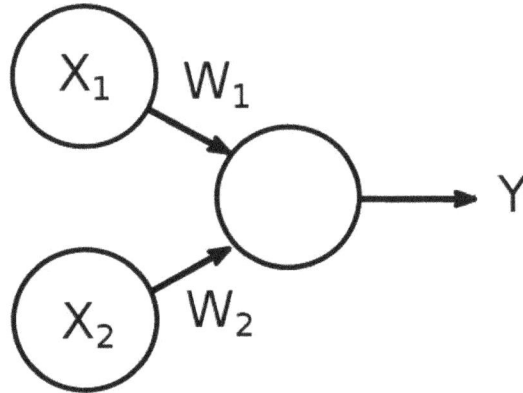

Figure A2. Sketch of a simple neuron.

Imagine now that, instead of a single input–output couple, P couples are available $(\mathbf{x}_1, y_1), \cdots, (\mathbf{x}_P, y_P)$, the learning can be expressed by minimizing the functional

$$\epsilon(\mathbf{W}) = \sum_{s=1}^{P} \left(y_s - \mathbf{W}^T \mathbf{x}_s \right)^2. \tag{A21}$$

The nonlinear case employs a nonlinear function of the predictor for the neuron activation. When the multicomponent inputs produces multicomponent outputs, **W** becomes a matrix instead the vector previously considered. However, the procedures for computing that matrix from the knowledge of P couples $(\mathbf{x}_s, \mathbf{y}_s)$, $s = 1, \ldots, P$ remain almost the same as the ones previously discussed. In some circumstances, instead of considering a single layer of neurons, multiple layers perform better.

Appendix A.6. Decision Trees

Decision tree learning is a technique that, given a data set $\mathcal{D} = \{((\mathbf{x}_i, y_i), i = 1, 2, \ldots, P\}$, where $\mathbf{x}_i$ are the so-called features and y_i the respective target variable to characterize, predicts an *outcome*, y for a new vector **x** that can be a class to which the set **x** belongs (in that case, we speak about *classification trees*), or a real number (in which case, we speak about *regression trees*).

In a regression tree, the terminal node j is assumed to have a constant value $y(j)$. The learning procedure consists of the top-down establishment of the variables that best partition the set according to a given criterion. One of the most popular criteria is the so-called Gini impurity criterion. For a set of M classes $m = 1, 2, \ldots, M$, and p_m the fraction of elements in $\mathcal{D}$ labeled within class m, the Gini impurity index is defined as

$$I(p) = \sum_{m=1}^{M} p_m \sum_{n \neq m} p_n$$

and measures the probability of an element to be chosen in its correct set times the probability of a wrong classification. In general, no matter the particular metrics employed, the goal is to classify the set so as to obtain as homogeneous as possible classification at terminal nodes.

Appendix B. Experimental Data

Table A1. Input parameters and training mask.

Training MaskIn	Screw Rotation Speed (rpm) INPUT	Exit Flow-Rate (kg/h) INPUT	Temperature (°C) INPUT	TS Dispersed Phase % INPUT	PP-g-MA/TS Ratio INPUT	Ratio TS INPUT
1	800	3	200	0	0	0.00
1	800	3	200	0	0	0.00
0	300	5	230	0	0	0.00
0	300	5	230	0	0	0.00
1	300	5	230	0	0	0.00
0	300	5	230	0	0	0.00
1	300	5	230	0	0	0.00
1	300	5	230	0	0	0.00
1	300	5	230	2	24.00	1.00
0	300	5	230	6	7.67	1.00
0	300	5	230	27	0.48	0.50
0	300	5	230	6	3.17	1.00
1	300	5	230	12	1.42	0.20
1	300	5	230	7	2.57	0.40
1	300	5	230	15	1.00	0.50
0	300	5	230	11	1.64	0.10
1	300	5	230	4	4.75	1.00
1	300	5	230	31	0.45	0.48
0	300	5	230	4	4.75	1.00
0	300	5	230	31	0.45	0.48
1	300	5	230	4	4.75	1.00
1	300	5	230	31	0.45	0.48
0	300	5	230	4	4.75	1.00
0	300	5	230	31	0.45	0.48
1	300	5	230	4	4.75	1.00
0	600	5	230	30	0.47	0.50
0	450	5	230	4	4.75	1.00
1	300	5	230	19	0.84	0.06
1	450	5	230	19	0.84	0.06
1	600	5	230	19	0.84	0.06
1	300	5	230	18	0.89	0.06
1	300	5	230	18	0.83	0.06
1	300	5	230	18	0	0.06

Table A1. Cont.

Training MaskIn	Screw Rotation Speed (rpm) INPUT	Exit Flow-Rate (kg/h) INPUT	Temperature (°C) INPUT	TS Dispersed Phase % INPUT	PP-g-MA/TS Ratio INPUT	Ratio TS INPUT
1	300	5	230	18	0.89	0.06
0	300	5	230	18	0.83	0.06
1	300	5	230	30	0.43	0.50
1	300	5	230	30	0	0.50
1	300	5	230	30	0.43	0.50
0	800	3	200	26	0.77	1.89
0	800	3	200	26	0.77	1.89
1	800	3	200	26	0.77	1.89
1	800	3	200	12	1.67	2.00
1	800	3	200	12	0.75	2.00
0	300	3	200	10	1.50	0.11
1	300	3	200	10	0.80	0.11
0	800	3	200	10	1.50	0.11
1	800	3	200	26	0.77	1.89
1	800	3	200	12	1.67	2.00
1	800	6	200	5.2	0.96	1.89
1	800	4.5	200	5.2	0.96	1.89
0	800	6	200	5.2	0.96	1.89
1	800	6	200	5	1.00	0.08
0	800	6	200	5	1.00	0.08
1	800	6	200	5	1.00	0.08
1	800	6	200	5.2	0.96	1.89
0	800	6	200	5.2	0.96	1.89
0	800	10	200	3.1	0.97	1.82
0	800	6	200	5	1.00	0.08
0	800	6	200	5	1.00	0.08

Table A2. Output quantities of interest.

Training MaskIn	E (MPa) OUTPUT	Yield (MPa) OUTPUT	σ_{break} (MPa) OUTPUT	ϵ_{break} OUTPUT	Re_{amb} (kJ/m) OUTPUT
1	1638	34	21	58	5
1	1505	37	26	164	5
0	1062	30	10	86	5.3
0	1357	21	21	10	4.9
1	1379	22	21	10	5.0
0	1281	20	13	69	7.6
1	1549	35	21	477	1.9
1	874	19	15	504	13.1
1	1345	29	14	106	5.6
0	1347	29	16	125	6.0
0	1673	30	21	118	7.4
0	1480	25	7	150	5.2
1	1657	28	12	201	6.1
1	1551	29	13	156	6.7
1	1723	30	12	155	5.8
0	1704	28	11	137	5.4
1	1365	21	19	22	4.3
1	1736	27	27	10	5.1
0	1284	22	15	136	5.6
0	1682	26	28	43	6.3
1	1416	34	24	606	2.4
1	1843	39	37	21	3.6
0	921	21	18	483	9.7
0	1371	26	24	49	12
1	1546	37	15	114	4.9
0	1753	39	29	63	5
0	1606	36	19	127	4.9
1	1655	38	16	105	3.8
1	1730	38	19	84	3.9
1	1802	38	21	54	3.3
1	1311	32	20	125	4.5
1	1181	30	18	180	5.8
1	1194	31	19	256	4.1
1	1073	24	14	155	3.7
0	1134	25	12	191	4.5

Table A2. Cont.

Training MaskIn	E (MPa) OUTPUT	Yield (MPa) OUTPUT	σ_{break} (MPa) OUTPUT	ϵ_{break} OUTPUT	Re_{amb} (kJ/m) OUTPUT
1	931	21	19	127	6.8
1	1180	25	16	63	4.4
1	1657	28	12	146	6.1
0	1795	40	37	10	5
0	2077	41	39	7	4.4
1	2273	41	39	10	3.4
1	1976	37	18	47	6.7
1	2021	36	23	44	6
0	1785	35	23	51	10.4
1	1939	37	26	38	10.1
0	1757	35	23	77	9.4
1	2033	38	23	44	5
1	2021	36	23	44	5
1	1943	37	17	46	5
1	2035	36	15	88	5
0	1897	36	26	88	5
1	1754	34	21	48	5
0	1807	35	17	49	5
1	1764	34	20	137	5
1	1067	20	17	215	5
0	2035	36	15	88	5
0	1825	37	16	51	5
0	1024	20	16	98	5
0	1807	35	17	49	5

Appendix C. Coefficients Involved in the Elastic Modulus Regression

Table A3. α_{ki}^1 coefficients.

k\i	0	1	2	3	4	5	6	7	8	9	10
0	1590.2	117.7	1.0	−1.0	−1.0	0.7	0.9	0.4	0.2	0.4	−0.3
1	0.0	171.6	0.0	0.0	0.0	−0.7	0.1	−0.6	1.0	−0.7	0.8
2	0.0	−91.9	0.0	0.0	0.0	−0.2	−0.5	0.7	0.1	0.6	−0.5

Table A4. α_{ki}^2 coefficients.

k\i	0	1	2	3	4	5	6	7	8	9	10
0	1.0	1.0	1.0	0.8	1.0	0.1	0.2	−0.6	0.5	0.0	−0.5
1	0.0	0.0	0.0	0.5	0.0	−0.7	−0.7	−0.1	−0.6	−1.0	0.0
2	0.0	0.0	0.0	0.3	0.0	0.7	0.7	0.8	0.6	0.1	−0.8

Table A5. α_{ki}^3 coefficients.

k\i	0	1	2	3	4	5	6	7	8	9	10
0	1.0	1.0	1.0	0.4	0.7	0.6	0.7	1.0	1.0	0.0	1.0
1	0.0	0.0	0.0	0.9	0.8	0.8	0.7	0.0	0.0	1.0	0.0
2	0.0	0.0	0.0	0.0	0.0	0.0	0.0	0.0	0.0	0.0	0.0

Table A6. α_{ki}^4 coefficients.

k\i	0	1	2	3	4	5	6	7	8	9	10
0	1.0	1.0	−0.2	1.0	0.4	1.0	1.0	1.0	1.0	1.0	1.0
1	0.0	0.0	0.5	0.0	0.9	0.0	0.0	0.0	0.0	0.0	0.0
2	0.0	0.0	−0.8	0.0	0.0	0.0	0.0	0.0	0.0	0.0	0.0

Table A7. α_{ki}^5 coefficients.

k\i	0	1	2	3	4	5	6	7	8	9	10
0	1.0	1.0	1.0	1.0	−0.7	−748.1	802.3	−0.6	0.1	1.0	0.5
1	0.0	0.0	0.0	0.0	−0.2	592.3	−159.1	−0.3	0.5	0.0	0.8
2	0.0	0.0	0.0	0.0	0.7	884.7	−917.2	0.7	0.9	0.0	0.4

Table A8. α_{ki}^6 coefficients.

k\i	0	1	2	3	4	5	6	7	8	9	10
0	1.0	1.0	152.2	104.3	66.3	1.0	1.0	235.1	−15.6	6.2	3958.4
1	0.0	0.0	57.1	51.5	−389.9	0.0	0.0	384.5	93.3	−10.4	9578.5
2	0.0	0.0	20.7	89.8	243.9	0.0	0.0	263.9	36.5	−11.5	4680.1

References

1. Chinesta, F.; Cueto, E.; Abisset-Chavanne, E.; Duval, J.L.; Khaldi, F.E. Virtual, Digital and Hybrid Twins: A New Paradigm in Data-Based Engineering and Engineered Data. *Arch. Comput. Methods Eng.* **2020**, *27*, 105–134. [CrossRef]
2. Ghnatios, C.; Alfaro, I.; Gonzalez, D.; Chinesta, F.; Cueto, E.E. Data-Driven GENERIC Modeling of Poroviscoelastic Materials. *Entropy* **2019**, *21*, 1165. [CrossRef]
3. González, D.; Chinesta, F.; Cueto, E. Thermodynamically consistent data-driven computational mechanics. *Contin. Mech. Thermodyn.* **2018**. [CrossRef]
4. González, D.; Chinesta, F.; Cueto, E. Learning corrections for hyperelastic models from data. Frontiers in Materials—section Computational Materials Science. *Front. Mater.* **2020**, in press.
5. Moya, B.; Gonzalez, D.; Alfaro, I.; Chinesta, F.; Cueto, E. Learning slosh dynamics by means of data. *Comput. Mech.* **2020**, in press.
6. Argerich, C.; Ibanez, R.; Barasinski, A.; Chinesta, F. Code2vect: An efficient heterogenous data classifier and nonlinear regression technique. *C. R. Mec.* **2019**, *347*, 754–761. [CrossRef]
7. Ibanez, R.; Abisset-Chavanne, E.; Ammar, A.; Gonzalez, D.; Cueto, E.; Huerta, A.; Duval, J.L.; Chinesta, F. A multi-dimensional data-driven sparse identification technique: The sparse Proper Generalized Decomposition. *Complexity* **2018**, 5608286.
8. Schmid, P.J. Dynamic mode decomposition of numerical and experimental data. *J. Fluid Mech.* **2010**, *656*, 5–28. [CrossRef]
9. Williams, M.O.; Kevrekidis, G.; Rowley, C.W. A data-driven approximation of the Koopman operator: Extending dynamic mode decomposition. *J. Nonlinear Sci.* **2015**, *25*, 1307–1346. [CrossRef]
10. Reille, A.; Hascoet, N.; Ghnatios, C.; Ammar, A.; Cueto, E.; Duval, J.L.; Chinesta, F.; Keunings, R. Incremental dynamic mode decomposition: A reduced-model learner operating at the low-data limit. *C. R. Mec.* **2019**, *347*, 780–792. [CrossRef]

11. Aguiar, L.G.; Pessoa-Filho, P.A.; Giudic, R. Modeling of the grafting of maleic anhydride onto poly(propylene): Model considering a heterogeneous medium. *Macromol. Theory Simul.* **2011**, *20*, 837–849. [CrossRef]
12. Beyer, G.; Hopmann, C. *Reactive Extrusion: Principles and Applications*; Wiley-VCH: Hoboken, NJ, USA, 2018.
13. Cassagnau, P.; Bounor-Legaré, V.; Fenouillot, F. Reactive processing of thermoplastic polymers: A review of the fundamental aspects. *Int. Polym. Process.* **2007**, *22*, 218–258. [CrossRef]
14. Cha, J.; White, J.L. Methyl methacrylate modification of polyolefin in a batch mixer and a twin-screw extruder experiment and kinetic model. *Polym. Eng. Sci.* **2003**, *43*, 1830–1840. [CrossRef]
15. Raquez, J.M.; Narayan, R.; Dubois, P. Recent advances in reactive extrusion processing of biodegradable polymer-based compositions. *Macromol. Mater. Eng.* **2008**, *293*, 447–470. [CrossRef]
16. Cassagnau, P.; Bounor-Legaré, V.; Vergnes, B. Experimental and Modeling aspects of the reactive extrusion process. *Mech. Ind.* **2020**, *20*, 820.
17. Hernandez-Ortiz, J.C.; Steenberge, P.H.M.V.; Duchateau, J.N.E.; Toloza, C.; Schreurs, F.; Reyniers, M.F.; Marin, G.B.; Dhooge, D.R. A two-phase stochastic model to describe mass transport and kinetics during reactive processing of polyolefins. *Chem. Eng. J.* **2019**, *377*, 119980. [CrossRef]
18. Vergnes, B.; Valle, G.D.; Delamare, L. A global computer software for polymer flows in corotating twin screw extruders. *Polym. Eng. Sci.* **1998**, *38*, 1781–1792. [CrossRef]
19. Vergnes, B.; Berzin, F. Modeling of reactive systems in twin screw extrusion: Challenges and applications. *C. R. Chim.* **2006**, *9*, 1409–1418. [CrossRef]
20. Choulak, S.; Couenne, F.; Gorrec, Y.L.; Jallut, C.; Michel, P.C.A. Generic dynamic model for simulation and control of reactive extrusion. *Ind. Eng. Chem. Res.* **2004**, *43*, 7373–7382. [CrossRef]
21. Goodfellow, I.; Bengio, Y.; Courville, A. *Deep Learning*; MIT Press: Cambridge, UK, 2016.
22. Cristianini, N.; Shawe-Taylor, J. *An Introduction to Support Vector Machines: In Addition, Other Kernel-Based Learning Methods*; Cambridge University Press: New York, NY, USA, 2000.
23. Garois, N.; Sonntag, P.; Martin, G.; Vatan, M.; Drouvroy, J. Procédé de Préparation D'une Composition Thermoplastique Renforcée et Réactive, et Cette Composition. 2014. Available online: https://patents.google.com/patent/EP2415824A1/fr (accessed on 12 June 2020).

© 2020 by the authors. Licensee MDPI, Basel, Switzerland. This article is an open access article distributed under the terms and conditions of the Creative Commons Attribution (CC BY) license (http://creativecommons.org/licenses/by/4.0/).

Article

Computational Predictions for Boger Fluids and Circular Contraction Flow under Various Aspect Ratios [†]

J. Esteban López-Aguilar [1,2,*] and Hamid R. Tamaddon-Jahromi [2]

1 Facultad de Química, Departamento de Ingeniería Química, Universidad Nacional Autónoma de México (UNAM), Ciudad Universitaria, Coyoacán, Ciudad de México 04510, Mexico
2 Institute of Non-Newtonian Fluid Mechanics, Zienkiewicz Centre for Computational Engineering, College of Engineering, Swansea University, Bay Campus, Fabian Way, Swansea SA1 8EN, Wales, UK; cshamid@swansea.ac.uk
* Correspondence: jelopezaguilar@unam.mx
† Dedicated to the Memory of Our Dear Friend and Colleague, Late Professor Mike Webster.

Received: 23 April 2020; Accepted: 27 May 2020; Published: 31 May 2020

Abstract: This work puts forward a modeling study contrasted against experimental, with focus on abrupt circular contraction flow of two highly-elastic constant shear-viscosity Boger fluids, i.e., a polyacrylamide dissolved in corn-syrup PAA/CS (Fluid-1) and a polyisobutylene dissolved in polybutene PIB/PB (Fluid-2), in various contraction-ratio geometries. Moreover, this work goes hand-in-hand with the counterpart matching of experimental pressure-drops observed in such 4:1 and 8:1 aspect-ratio contraction flows, as described experimentally in the literature. In this study, the experimental findings, for Boger fluids with severe strain-hardening features, reveal significant vortex-evolution characteristics, correlated with enhanced pressure-drop phasing and normal-stress response in the corner region. It is shown how such behavior may be replicated through simulation and the rheological dependencies that are necessary to bring this about. Predictive solutions with an advanced hybrid finite-element/volume (*fe/fv*) algorithm are able to elucidate the rheological properties (extensional viscosity and normal-stress response) that rule such vortex-enhancement evolution. This is accomplished by employing the novel swanINNFM(q) family of fluids, through the swIM model-variant, with its strong and efficient control on elongational properties.

Keywords: Boger fluids; circular contraction flow; lip vortex; pressure-drops; vortex-enhancement; first normal-stress difference; swIM model

1. Background and Introduction

Quantitative comparison between numerical predictions, experimental observations, and complex flow, occurring in contraction and contraction–expansion flows, has occupied the attention of the rheological scientific community over decades; see, for instance, Boger [1] and Boger et al. [2], López-Aguilar et al. [3], Tamaddon-Jahromi et al. [4], Nigen and Walters [5], Binding et al. [6], Pérez-Camacho et al. [7], Tamaddon-Jahromi et al. [8], López-Aguilar et al. [9], López-Aguilar et al. [10], Webster et al. [11], and reviews of Walters and Webster [12], White et al. [13], and Owens and Phillips [14]. This may be recorded in terms of pressure-drops, and vortex-activity in the recess-zones nearby salient and re-entrant corners of these geometries. In Boger [1] and Boger et al. [2], attention was given to two experimental studies with highly-elastic constant-viscosity 'Boger' fluids and circular contractions. In the first of these two studies [2], comparison was made between the flow of two Boger fluids, with basically the same principal characteristic relaxation-times, in three contraction-ratio geometries (α_{aspect}) of 2:1, 4:1, and 16:1. Findings with increasing shear-rate disclosed two distinct kinematical patterns. The first fluid under α_{aspect} = {4, 16}, a polyacrylamide in corn-syrup PAA/CS solution (Fluid-1, Figure 1), showed continual salient-corner vortex-growth, with separation-line adjustment in shape from concave-to-convex. In contrast, the second test polyisobutylene in polybutene PIB/PB solution (Fluid-2, Figure 2) displayed a sequential combination of salient-corner and lip vortices. Then, as deformation-rate increased, the lip kinematical structure completely engulfed the shrinking salient-corner vortex, giving way to a single large recirculating entity of convex shape (elastic-corner vortex). Note, in both experiments and at high shear-rates, the vortex extended in coverage up to the re-entrant corner. Under such different scenarios, Boger et al. [2] concluded that measurement of steady and dynamic shear properties alone were insufficient to characterize the response of such elastic liquids in circular contraction flow.

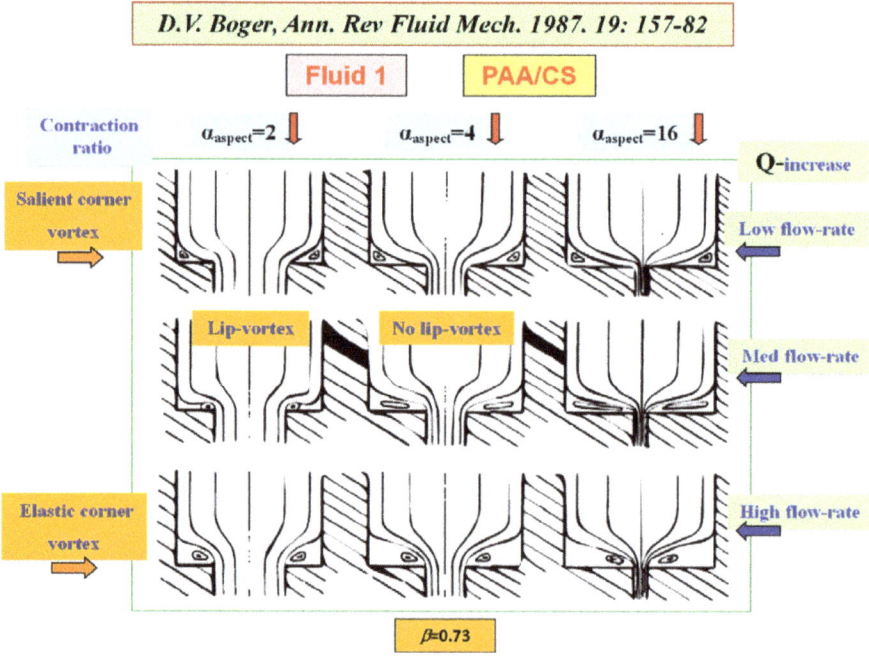

Figure 1. Vortex-activity with increasing flow-rate Q; Boger Fluid-1 (PAA/CS); β = 0.73, α_{aspect} = {2, 4, 16}.

In the second and further study, Boger and Binnington [15] studied two Boger fluids to produce streak-like photographic observations for α_{aspect} = 4 circular contractions, of sharp and rounded-corner configurations. The first fluid was the organically-based international test fluid Ml, (polyisobutylene dissolved in polybutene (PIB/PB)). The second fluid was referred to as fluid P1 (0.03% polyacrylamide dissolved in corn-syrup (PAA/CS)). Both fluids exhibited significant elasticity, while at the same time, a constant shear viscosity. There, these two fluids exhibited distinctly different vortex-enhancement paths for a given aspect-ratio choice. For example, in the rounded-corner geometry, the salient-corner vortex appeared almost constant in size with the M1-fluid, whilst lip-vortex formation was observed under the P1 fluid case. As for the abrupt contraction, the M1-fluid displayed marginal vortex-growth, whilst vortex-enhancement was more active for the P1-fluid. Once more, these major differences between the responses in complex-flow of solutions with similar shear properties in fixed geometries render their extensional features as the subjacent explanation for such diversity of trends in vortex-activity.

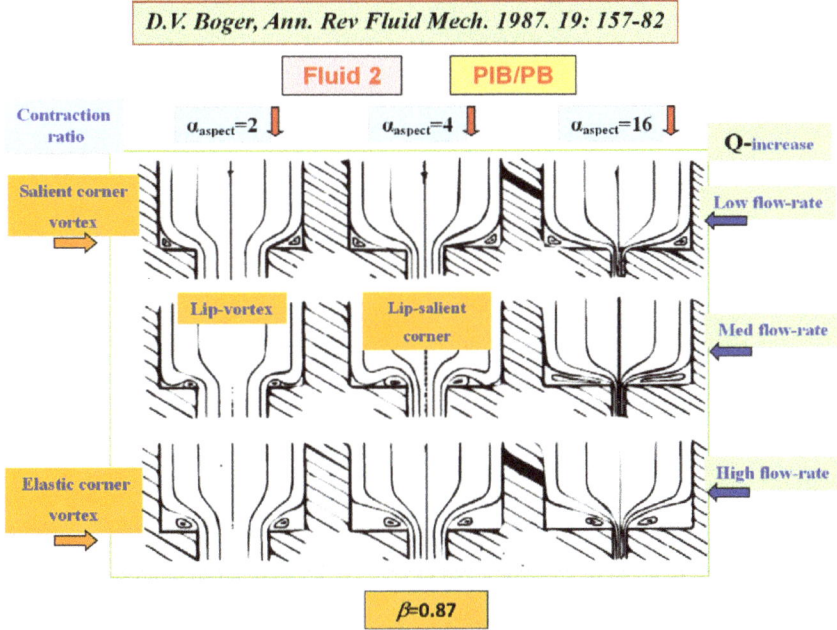

Figure 2. Vortex-activity with increasing flow-rate Q; Boger Fluid-2 (PIB/PB); $\beta = 0.87$, $\alpha_{aspect} = \{2, 4, 16\}$.

In addition, Rothstein and McKinley [16] explored experimentally the creeping flow of a dilute 0.025 wt% polystyrene/polystyrene (PS/PS) Boger fluid. These authors, covering a large range of Deborah numbers, devoted attention to circular contraction–expansion flow-settings of various aspect-ratios ($\alpha_{aspect} = \{2, 4, 8\}$) and re-entrant corner curvature degrees. For a relatively low aspect-ratio of $\alpha_{aspect} = 2$ of sharp-corners, a steady lip-vortex was observed at the re-entrant corner. For aspect-ratios of between $\alpha_{aspect} = 4$ and 8, lip-vortex formation was absent, but a salient-corner vortex was recorded, which grew with the flow-rate increase. Rounding the re-entrant corner shifted such landmarks and trends to higher values of flow-rates, but did not change qualitatively the structure and evolution of the overall flow-field.

Sato and Richardson [17] performed simulations for planar $\alpha_{aspect} = 4$ contraction flow. These were based on a hybrid finite volume/element method, embedded in a time-stepping procedure within a pressure-correction scheme. These authors reported lip-vortex formation as a pseudo-transient phenomenon, appearing at Reynolds-number levels of $Re = 0.01$, and being triggered by an instantaneous increase in Deborah numbers (De) from six to twelve. Subsequently, such transient lip-vortex faded through the time-stepping process, as a steady-state solution was approached at the limiting value of elasticity of $De = 12$. Similarly, for the same $\alpha_{aspect} = 4$ planar contraction flow, Olsson [18] also observed the transient presence of a lip-vortex, but using the Giesekus rheological equation-of-state; whilst employing a method-of-lines technique for time-integration and a discretization based on finite-differences.

In keeping with the above developments, the present study considers counterpart predictive solutions generated with a hybrid-subcell finite-element/volume algorithm (fe/fv) [19–21], incorporating some novel advanced stabilization techniques [22,23]. Attention is directed towards contraction-ratios of $\alpha_{aspect} = \{2, 4, 8\}$, covering in particular the correlation of pressure-drop enhancement, vortex-dynamics (lip-vortex formation), and flow-structure (normal-stress response); the context is one of Boger fluids and creeping flow conditions. This range of contraction-ratios was held sufficient for present comparison purposes, as gathered from our prior work on contraction-expansion ratio comparison in

López-Aguilar et al. [9]. An appeal is also made to our companion study in Tamaddon-Jahromi et al. [4], where the focus of attention there was solely on the $\alpha_{aspect} = 8$ contraction-ratio problem.

2. Governing Flow Equations, Material Functions, Problem Specification, and Numerical Algorithm

Following the principles of conservation of momentum and mass, the non-dimensional equations that govern the flow response of viscoelastic fluids under creeping incompressible and isothermal conditions, are:

$$\nabla \cdot \boldsymbol{u} = 0 \tag{1}$$

$$Re\frac{\partial \boldsymbol{u}}{\partial t} = \nabla \cdot \boldsymbol{T} - Re\, \boldsymbol{u} \cdot \nabla \boldsymbol{u} - \nabla p. \tag{2}$$

As such, a domain bounded in space and time (x, t) is considered, over which spatial–temporal differential operators apply. Then, field variables $\boldsymbol{u}$, p, and $\boldsymbol{T}$ represent fluid velocity, hydrodynamic pressure, and stress-tensor, respectively. The stress-tensor may be expressed as:

$$\boldsymbol{T} = \boldsymbol{\tau} + 2\beta \boldsymbol{d}. \tag{3}$$

The stress-tensor is decomposed into two parts by means of the Elastico-Viscous Stress Splitting (EVSS) assumption, where the total stress $\boldsymbol{T}$ is composed by two contributions, one for the polymer, to which the viscoelastic nature is addressed through $\boldsymbol{\tau}$, and another of Newtonian-like response of the form $2\beta \boldsymbol{d}$. In this, $\boldsymbol{d} = (\nabla \boldsymbol{u} + \nabla \boldsymbol{u}^{\dagger})/2$ represents the rate-of-deformation tensor, where tensor transpose is denoted with the superscript †. In addition, the non-dimensional group Reynolds number is defined as $Re = \frac{\rho U_{char} L_{char}}{\mu_0}$, through characteristic scales of U_{char} on fluid velocity (mean flow-velocity over the characteristic-length), and, for length, L_{char}, as the constriction radius. The material density is represented with ρ and the characteristic viscosity taken as a zero shear-rate viscosity ($\mu_0 = \mu_p + \mu_s$). Here, μ_p and μ_s are the polymeric viscosity and the solvent viscosity components, respectively, so that the solvent-fraction parameter can be defined as $\beta = \frac{\mu_s}{\mu_0}$. Creeping flow conditions are presumed throughout, so that Reynolds numbers are typically $O(10^{-2})$ or smaller.

2.1. Constitutive Modeling

To complete the equation set, one needs a state law on stress, which is provided by the swanINNFM(q) model formalism (see Tamaddon-Jahromi et al. [4,8] and López-Aguilar et al. [9]). This model is soundly-based, being derived from two well-respected models, the Finitely Extensible Non-linear Elastic dumbbell Chilcott-Rallison FENE-CR model (Chilcott and Rallison [24]) and a White–Metzner model (White and Metzner [25]).

The base FENE-CR model may be written in a configuration-tensor A form (bold-face symbols denote tensorial quantities), as:

$$Wi\, \overset{\nabla}{\boldsymbol{A}} + f[tr(\boldsymbol{A})](\boldsymbol{A} - \boldsymbol{I}) = 0. \tag{4}$$

Here, $\overset{\nabla}{\boldsymbol{A}}$ stands for the upper-convected material-derivative of the configuration-tensor ($\boldsymbol{A}$), defined as:

$$\overset{\nabla}{\boldsymbol{A}} = \frac{\partial \boldsymbol{A}}{\partial t} + \boldsymbol{u} \cdot \nabla \boldsymbol{A} - (\nabla \boldsymbol{u})^{\dagger} \cdot \boldsymbol{A} - \boldsymbol{A} \cdot (\nabla \boldsymbol{u}). \tag{5}$$

The FENE-CR structural-functional $f[tr(\boldsymbol{A})]$ is:

$$f[tr(\boldsymbol{A})] = \frac{1}{1 - \frac{tr(\boldsymbol{A})}{L^2}}. \tag{6}$$

Then, Kramer's rule relates configuration and extra-stress tensors as follows:

$$\tau = \frac{1-\beta}{Wi} f[tr(A)](A - I). \tag{7}$$

In the above, L is the extensibility parameter for the FENE-CR model, related to the dumbbell chain-length, and I is the identity tensor. In addition, the non-dimensional Weissenberg group-number is defined as $Wi = \lambda_1 \frac{U_{char}}{L_{char}}$, where dependency upon the fluid relaxation-time (λ_1) and a characteristic rate ($\frac{U_{char}}{L_{char}}$) is observed. Then, rise in Wi may be generated through deformation-rate increase, fixing the fluid elastic character through λ_1, whilst the flow-rate Q-dynamics are increased, i.e., $Wi = \lambda_1 \frac{Q}{\pi L_{char}^3}$, considering that $Q = AU_{char} = \pi L_{char}^2 U_{char}$.

To arrive at the swanINNFM(q) model, one needs to consider a rate-dependent viscosity in the above developments, as under the generalized White–Metzner model, taking this to be extension-rate dependent alone. Then, using Equations (3) and (7) above, the new resulting family of swanINNFM(q) models, in its single relaxation-time swIM model-variant [3,4,8–11], may be articulated through the amended total-stress tensor, as:

$$T = \frac{1-\beta}{Wi} f[tr(A)](A - I)\phi(\dot{\varepsilon}) + 2\beta\phi(\dot{\varepsilon})d, \tag{8}$$

where the dissipative extensional-function $\phi(\dot{\varepsilon})$ is taken as a quadratic-form from the truncated Taylor-series approximation of the cosh-exponential expression available. This dissipative extensional-function is defined as $\phi(\dot{\varepsilon}) = 1 + (\lambda_D \dot{\varepsilon})^2$, with parameterization on a dissipative material time-scale parameter λ_D, and functionality on a generalized strain-rate invariant $\dot{\varepsilon} = \frac{III_d}{II_d}$. Here, III_d and II_d represent the third and the second invariants of d, respectively. Fuller details on the development of this swanINNFM(q)-family of fluids are supplied in Debbaut and Crochet [26], Debbaut et al. [27], Tamaddon-Jahromi et al. [4,8], López-Aguilar et al. [3,9,10], and Webster et al. [11].

2.2. Material Functions

The relevant swIM-model rheometrical functions are provided in Figure 3, noting a constant shear-viscosity. These are the extensional viscosity η_e and the first normal-stress difference in shear N_{1Shear}, where variation over model parameters (β, L, λ_D) is presented. Their functional forms are, respectively:

$$\eta_e = 3\phi(\dot{\varepsilon})\beta + 3\phi(\dot{\varepsilon})(1-\beta)\left[\frac{f^2}{f^2 - fWi\dot{\varepsilon} - 2Wi^2\dot{\varepsilon}^2}\right], \tag{9}$$

$$N_{1Shear} = \frac{2(1-\beta)Wi\dot{\gamma}^2}{f}. \tag{10}$$

In Figure 3a, a solvent-content β-variation extensional viscosity η_e-response is exposed for the swIM model, under $\lambda_D = 0.075$, $L = 5$, and $\beta = \{0.9, 0.8, 0.5, 1/9\}$. Firstly, the swIM model-response appears bounded by the two extremes of its behavior; under $\lambda_D = 0$, the dissipative extensional influence disappears, and the swIM model extensional hardening finds a plateau for moderate-to-high extension-rates; on the other extreme, at $L \to \infty$, an Oldroyd-B-like response is recovered, with infinite extensional viscosity predictions. Within these two bounds, swIM β-decrease, which may be interpreted as an increase in solute-content, renders a rise in the plateau-level observed at intermediate shear-rates in the range of $1 \leq \lambda_1 \dot{\varepsilon} \leq 70$ units. Beyond such plateaued-stage, a steep rise is witnessed as a result of the influence of the extensionally-driven dissipative mechanism promoted by $\phi(\dot{\varepsilon}) = 1 + (\lambda_D \dot{\varepsilon})^2$; here, β-decrease shifts this η_e-rise to higher extension-rates. In Figure 3b, the first normal-stress in shear N_{1Shear}-response is plotted for both Oldroyd-B and swIM under β-decrease. Here, in contrast to the stiff quadratic rising Oldroyd-B N_{1Shear}-trend, swIM provides a softer trend for shear-rates beyond

$\lambda_1 \dot\gamma$~10 units; in addition, β-decrease shifts N_{1Shear}-rise to lower shear-rates. Such predictive capabilities of the swIM model are contrasted against N_{1Shear} data reported by Boger et al. [2] for both Boger fluids formed by diluted solutions of PAA/CS and PIB/PB. This data is presented in dimensionless form, taking as characteristics time and viscosity scales, the characteristic time and the constant shear-viscosity reported experimentally by Boger et al. [2] as $\{\lambda_1, \eta_0\} = \{0.380\ s, 97.5\ P\}$ for the PAA/CS solutions and $\{\lambda_1, \eta_0\} = \{0.149\ s, 251\ P\}$ for the PIB/PB case. Stark matching is recorded between the experimental rheometrical N_{1Shear} and the predictions achieved using the swIM model. Particular to Figure 3b, a window of experimental-data capture is defined in the ranges of solvent-fraction $0.5 \leq \beta \leq 0.9$, extensibility-parameter $5 \leq L \leq 12$, and dissipative-parameter $0 \leq \lambda_D \leq 0.1$, which is used in subsequent sections for the simulation of contraction complex flow of those Boger fluids.

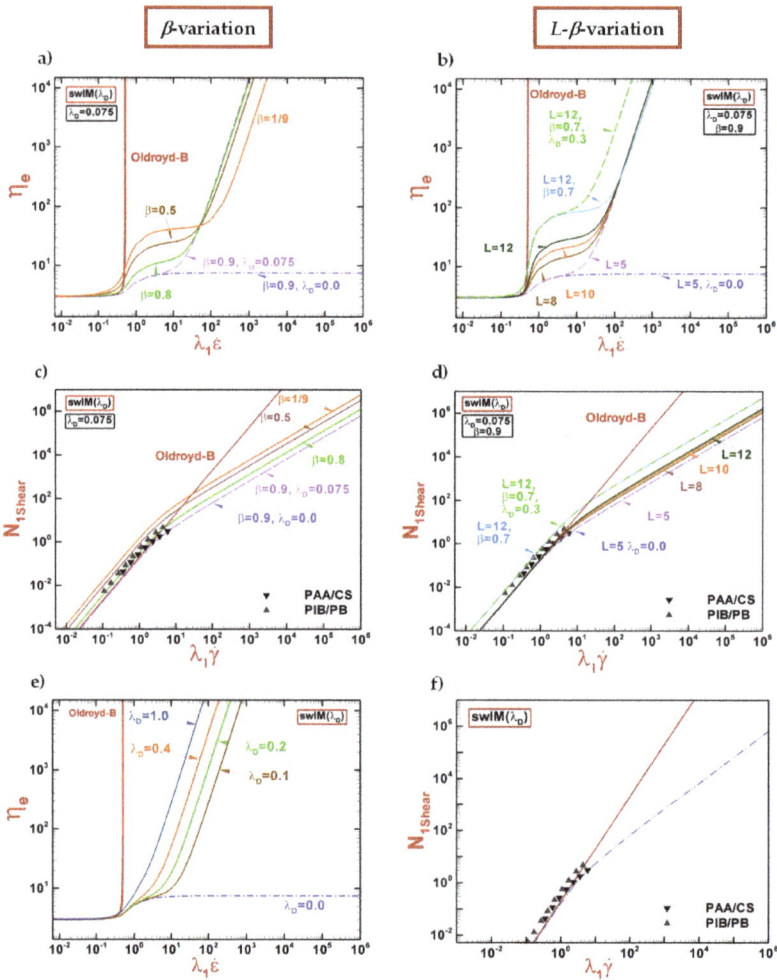

Figure 3. (**a,c**) Extensional viscosity η_e; (**b,d**) first normal-stress difference in shear N_{1Shear}; $\lambda_D = [0.0, 0.075]$; (**a,b**) $1/9 \leq \beta \leq 0.9$, $L = 5$; (**c,d**) $5 \leq L \leq 10$, $\beta = \{0.9, 0.7\}$; (**e**) Extensional viscosity η_e; (**f**) first normal-stress difference in shear N_{1Shear}; $\lambda_D = [0, 0.1, 0.2, 0.4, 1]$, $\beta = 0.9$, $L = 5$; Oldroyd-B and swIM models; symbols: experimental N_{1Shear} data from Boger et al. [2].

In Figure 3c, swIM extensional viscosity response with L-variation is reported. Particularly, this parametric study is performed under $\beta = \{0.9, 0.7\}$, $\lambda_D = \{0, 0.075\}$, and $L = \{5, 8, 10, 12\}$. swIM extensional viscosity under L-increase is analogous to that observed under solvent-fraction β-decrease, displaying an intermediate plateaued region, and followed by a sharp increase. Interestingly, the cumulative response of increasing both solvent-fraction β, the extensibility-parameter L and the extensional-dissipative time-scale λ_D of $\beta = 0.7$, $L = 12$ and $\lambda_D = 0.3$, exposes the strength of this model to boost hardening in extensional viscosity. The effects of such variations in first normal-stress in shear N_{1Shear} are, under L-increase, to enhance elasticity beyond $\lambda_1 \dot{\gamma} \sim 10$ units (Figure 3d). Notably, the coincidence between swIM predictions and N_{1Shear} experimental data-trends reported by Boger et al. [2] holds.

In Figure 3e, the influence of λ_D-variation over extensional viscosity is provided under $\beta = 0.9$ and $L = 5$. Here, the departure from the FENE-CR trend at intermediate extensional-rates appears sooner as the level of λ_D is larger, even vanishing the plateaued section for $\lambda_D = 0.4$ in comparison with smaller λ_D-cases. In terms of first normal-stress in shear N_{1Shear} (Figure 3f), as devised for this swIM model, λ_D-increase does not affect response in shear deformations.

2.3. Problem Specification and Numerical Scheme

The meshes used to discretize the problem, described in a 2D-domain, under three aspect-ratios $\alpha_{aspect} = \{2, 4, 8\}$, are displayed in Figure 4. Their characteristics in terms of node number and degrees-of-freedom are provided in Table 1. On mesh-refinement and solution-convergence, one may refer to the counterpart study in Tamaddon-Jahromi et al. [4], where this topic is well-covered. One may note the additional fine meshing used in the $\alpha_{aspect} = 4$ instance, around the re-entrant corner zone, which is necessary to pursue a stringent lip-vortex search.

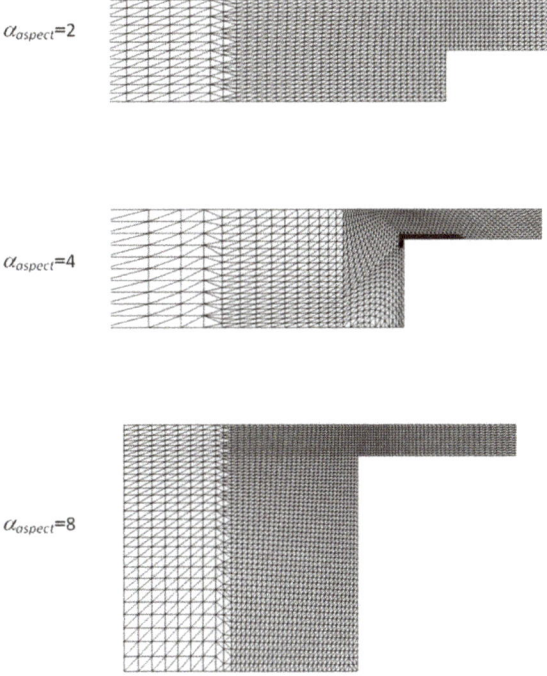

Figure 4. Zoomed mesh sections of contraction geometries; $\alpha_{aspect} = \{2, 4, 8\}$.

Table 1. Mesh characteristics, different contraction aspect-ratios.

Mesh α_{aspect}	Elements	Nodes	Degrees of Freedom (u,p,τ)
2	2762	5787	36,235
4	2987	6220	38,937
8 (refined)	2016	4191	26,234
8 (medium)	1707	3634	22,768
8 (coarse)	868	1897	11,897

Boundary conditions. On flow boundary-conditions, the flow specification is as follows. No-slip is taken on boundary walls, and shear-free symmetry is imposed on the flow-centerline (see also below). At flow-inlet, velocity and stress are specified, according to the flow-rate setting, akin to full-developed shear-flow. There, under vanishing convective terms, the partial differential equations PDE for stress-configuration collapses to a temporal ordinary differential equation ODE system, providing evolution to the algebraic stress equivalent forms. At outlet, only the arbitrary level of pressure is set. Then, through a steady-state solution-continuation procedure, initial conditions from a prior flow-rate solution may be accessed, and under more severe-flow parameter selection, feed-forward exit-procedures may be used on velocity-gradients and stress components to accelerate convergence (see López-Aguilar et al. [9,22]). Such feed-forward procedure overwrites fully-developed polymeric-stress τ and velocity-gradient ∇u components from inter-field regions towards the outlet-edge neighborhood prior solution approximation in each time-step. This helps to reduce noise proliferation originated at the outlet, which reflects back towards the internal field, and is particularly useful under moderate-to-high flow-rates (López-Aguilar et al. [9,22]). Generally, a flow-rate increase mode is adopted through a series of steady-state solutions, as appropriate and as prescribed elsewhere (Tamaddon-Jahromi et al. [4,8], López-Aguilar et al. [3,9,10]).

On numerical-to-experimental scaling. In the $\alpha_{aspect} = 4.08$ contraction flow of Boger [1] and Boger et al. [2], a Weissenberg number definition was introduced as $Wi^{exp} = \lambda^{exp}\dot{\gamma}$; where λ^{exp} is a relaxation-time and $\dot{\gamma}$ is a characteristic shear-rate (downstream wall shear-rate in tubular entry flow). Accordingly, respective relaxation-times were identified of $\lambda^{exp} = 0.149$ s for the polyisobutylene (PIB) in polybutene (PB) Boger fluid, and $\lambda^{exp} = 0.308$ s for the polyacrylamide (PAA) in water and corn syrup (CS) Boger solution. A functional relationship may be derived between these two experimental and computational Wi-definitions; experimentally of ($Wi^{exp} = \lambda^{exp}\dot{\gamma}$) and computationally of ($Wi^{comp} = \lambda_1 \frac{U_{char}}{L_{char}}$). For one-to-one comparison purposes, this establishes appropriate scaling factors of ($\frac{1}{\lambda^{exp}}$), between experimental findings and computational predictions per fluid-instance; yielding: $Wi^{comp} = 6.71\, Wi^{exp}$ for fluid PIB/PB and $Wi^{comp} = 3.25 Wi^{exp}$ for fluid PAA/CS (see López-Aguilar et al. [3] and Tamaddon-Jahromi et al. [4] for more details on such scaling). Based on these scaling factors, Wi^{exp} as in Boger [1] compares as $10.5 \leq Wi^{comp} \leq 16.6$ under fluid PIB/PB, and $2.05 \leq Wi^{comp} \leq 5.3$ under fluid PAA/CS. In practice, one notes below for the $\alpha_{aspect} = 4$, that computationally, a slight lip-vortex appears for $2.5 \leq Wi^{comp} \leq 5.5$ (lower rate range); whilst experimentally, a lip-vortex is only observed for fluid PIB/PB (in the higher Wi^{comp}-range) and not recorded with fluid PAA/CS.

Hybrid subcell finite-element/finite-volume scheme. The numerical method used in this work is based on a hybridized finite-element (*fe*) and finite-volume (*fv*) spatial-discretization scheme. Such a formulation comprises both time-stepping and fractional-staged (three) equation-structure. On the momentum-mass conservation equation doublet, finite-element (*fe*) discretization, grafted upon a Taylor–Petrov–Galerkin structure, is selected following incremental pressure-correction strategy. On the constitutive stress-equation, finite-volume (*fv*) discretization is employed. Such a space-time discretization agrees with equation-type specification. Accordingly, Galerkin-type (*fe*) approximation is applied on parent triangular tesselations; whilst a subtended subcell/cell-vertex finite-volume (*fv*) discretization is used for the rheological equation-of-state on stress-tensor components. An element-by-element iterative solution-procedure, space-efficient in its implementation, is utilized for discretized equations.

The pressure-equation is solved with a direct Choleski-reduction method. Then, the (*fv*)-component on stress is treated in a direct single-iteration implementation. The conservation-form for the stress equation is non-linear, and contains inhomogeneous source terms. This demands both median-dual-cell treatment for source terms and fluctuation-distribution for fluxes (upwinding). Additionally, quadratic interpolation is chosen for velocity, whilst linear interpolation is specified for pressure on the parent *fe* triangular-cell grid. For the finite-volume implementation, four *fv*-subcells per parent *fe*-cell are obtained, being the *fv*-sub-cell constructed via the interconnection of the parent *fe*-cells mid-side nodes. In such a structured arrangement, stress variables are located at the vertices of *fv*-sub-cells, and solution projection between is unnecessary. On the child subcell-level, this provides for a subcell-vertex *fv*-method for which trial-solutions are interpolated linearly. The resulting formulation is consistent in time and holds an accuracy of second-order. Further details on the numerical scheme and its detailed implementation characteristics, can be found in Wapperom and Webster [19], Webster et al. [20], and Aboubacar and Webster [21].

Stabilization techniques. Additional and latest aspects of improved stabilization techniques for viscoelastic flow employed are summarized as follows. The set of such stabilization techniques comprises the velocity-gradient VGR-correction [22,23], the use of velocity-gradient recovery, a discrete continuity correction over the flow-domain, and additional compatibilizing conditions on the flow-centerline, with pure-extension shear-free inhomogeneous extensional deformation at the centerline symmetry. Additionally, the absolute-value ABS-*f* correction [22,23] regularizes the problem through absolute-value imposition on the structure-network functional (*f*) in the rheological equation-of-state and within the Kramer's rule transformation in the momentum equation. Positive-definiteness of the problem is promoted through the use of configuration-tensor form in the constitutive equation [3,4,8,9,11,22,23].

3. Results—Computational Predictions and Flow-Structure versus Pressure-Drop Correlation

Comparison of predictive solutions is presented across the three geometric aspect-ratios, in turn, of $\alpha_{aspect} = \{2, 4, 8\}$. In this, it is informative to consider ramping-up through flow-rate (low, medium, high), where, due to the variation in dynamics per geometry, these ranges themselves will vary per geometry. Specific insight is drawn through comparison against the counterpart experimental patterns of Boger [1] and Boger et al. [2], and particularly, when focusing on Boger fluids of two different solvent-fractions of $\beta = 0.87$ and $\beta = 0.73$, as extracted above in N_{1Shear}-match with swIM model in Section 2.2. The Results section is organized in two main subsections. Firstly, 3.1 8:1 contraction flow includes: flow-structure (vortices and first normal-stress difference N_1) and pressure drops, renders the main findings of this work, and evidences the matching of experimental pressure-drops using the swIM model. Here, conspicuously, vortex-development phasing is correlated with pressure-drop enhancement, as described by Binding and Walters [28], providing theoretical explanation to experimental features of this benchmark circular contraction problem. Moreover, viscoelastic response in the recess-zones, observed through first normal-stress N_1-fields, appears directly linked with the vortex-formation and evolution, with salient-corner, lip, and elastic corner vortex capture, where the tracking of the shape and size of vortex-structures is recorded. This lies as a major finding on relating pressure-drop enhancement, vortex-evolution (flow kinematics), and flow structure [9]. The second section, Section 3.2 Predictive capabilities of the swIM—Vortex-dynamics across $\alpha_{aspect} = \{2, 4, 8\}$ circular contraction flow, provides insight into the influential swIM model parameters that permit the prediction of the elusive lip-vortex. Particular attention is paid to swIM solvent-fraction β, extensibility-parameter L and dissipative parameter λ_D variation, for which exploration of their vortex-dynamics and lip-vortex-formation are explored, and where intermediate λ_D-values at relatively high extensibility L-features appear as a proper combination for lip-vortex capture. Such parametrical-specification correlates with the precise control of extensional properties provided by the swanINNFM(q) model-family, embodied here through its swIM variant.

3.1. 8:1 Contraction Flow: Flow-Structure (Vortices and First Normal-Stress Difference N_1) and Pressure Drops

Notably and overall, all geometric aspect-ratios explored display elastic-corner vortices (*ecv*) at large flow-rates, as it is apparent in Figures 1 and 2. Hence, to describe the evolution of such kinematic structures and its relationship with pressure-drops with flow-rate rise, one may begin with the $\alpha_{aspect} = 8$ aspect-ratio case (see Tamaddon-Jahromi et al. [4]), as this instance provides the strongest dynamics and the sharpest distinction in flow-pattern features and viscoelastic pressure-drop rise arising experimentally. Such an $\alpha_{aspect} = 8$ geometry is reflective of higher ratios and reveals the evolving streamline patterns and exaggerated pressure-drop trends of Figure 5. First, recorded in relatively low flow-rate range of $Q/Q_0 \leq 0.035$ units, salient-corner vortices (*scv*) at low-rates arise, accompanied by pressure-drops that concur with Newtonian equivalents (here, Q_0 and Δp_0 are, respectively, characteristic flow-rate and pressure-drop taken from experiments [1]). Then, with flow-rate increase, in the low-to-mid flow-rate range of $0.12 \leq Q/Q_0 \leq 0.2$ units, such relatively simple behavior gives way to co-existent salient-corner/lip vortices (*lv*), instance for which viscoelastic pressure-drop initiates its departure from Newtonian response. Further flow-rate increase drives coalescence of the co-existent *scv* and *lv*, and marks the entry to the onset of large elastic-corner vortices (*ecv*) in the mid-to-high flow-rate regime of $Q/Q_0 \geq 0.35$ units, in which stark departure in pressure-drop between simple Newtonian and viscoelastic Boger fluids is apparent. Such vortex-evolution, from salient-corner vortex to elastic-corner vortex, has been proposed by Binding and Walters [28] as a cause for the pressure-drop enhancement observed experimentally.

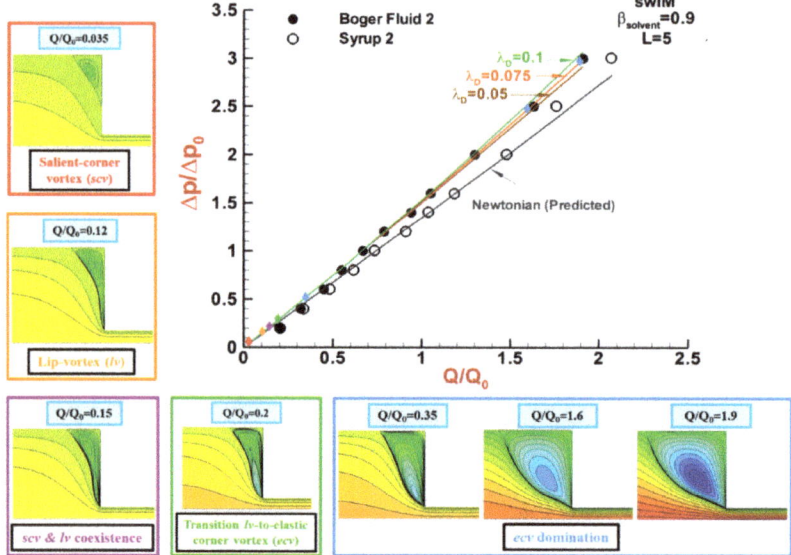

Figure 5. Pressure and streamlines against flow-rate; swIM model; $\lambda_D = \{0.0, 0.075, 0.1\}$; $\beta \leq 0.9$, $L = 5$. Note the coloring of the stages under each stream-line pattern, indicating the vortex-type present at each flow-rate and its place in the pressure drop plot. Here, a vortex-evolution is depicted, from salient-corner vortices, passing lip-vortex generation, its co-existence with the salient-corner kinematic structure, followed by elastic-corner vortex domination.

Correlation of complex flow features and Boger-fluid normal-stress difference in complex flow. It was established in López-Aguilar et al. [9] that the various vortex-structures and flow-stages (*scv*, *lv*, and *ecv*) tie in closely with the corner-patterns sustained in N_1-fields from complex flow, whilst vortex-evolution with Q-increase across such *scv*, *lv*, and *ecv* stages are closely driven the extensional viscosity response in ideal extensional deformation, as predicted with the swIM-model. The present study would concur with this, as evidenced in Figure 6, where sample streamlines are contrasted against N_1-fields for $\alpha_{aspect} = 8$ case. Note that in the circular contraction complex flow at hand, the first normal-stress difference is defined as $N_1 = \tau_{zz} - \tau_{rr}$, where τ_{zz} and τ_{rr} represent the normal-stresses in z and r directions, respectively. Here, the phases of *scv* growth, co-existence of *scv-lv*, and *ecv*-domination are mirrored in the N_1 field-data of Figure 6. Under this evidence, one may conclude that elasticity and, hence, non-linearity, manifest through first normal-stress difference in shear and extension (extensional viscosity), is observed to strongly influence the formation of such vortex-structures and its counterpart energetic pressure-drop effects in departure from Newtonian equivalent levels.

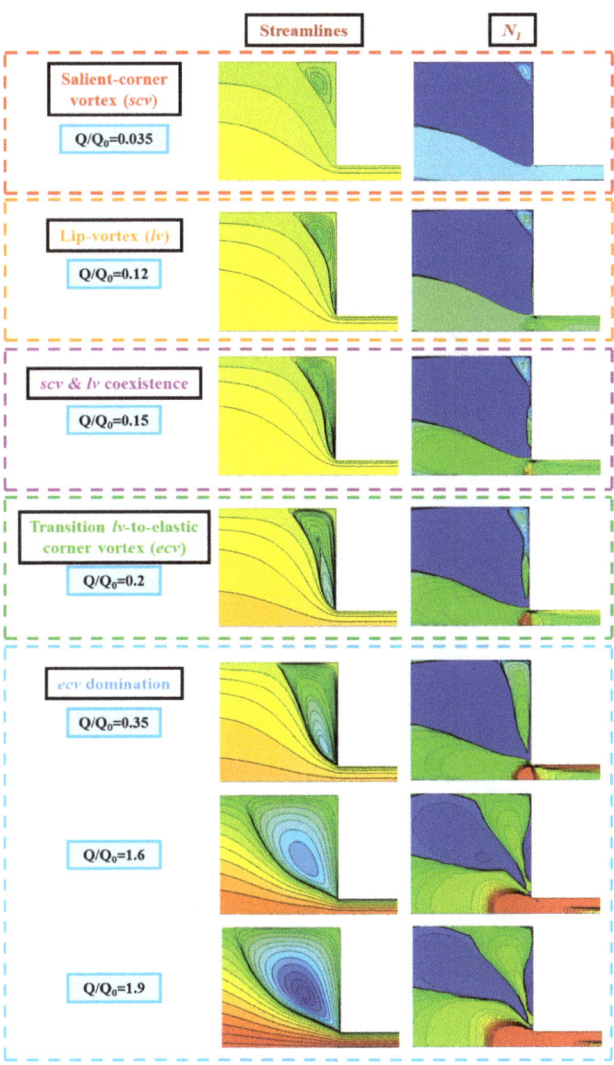

Figure 6. Streamlines and N_1 fields against flow-rate; swIM model; $\lambda_D = \{0.075\}$; $\beta \leq 0.9$, $L = 5$.

3.2. Lip-Vortex Predictive Capabilities of the SwIM—Vortex-Dynamics across $\alpha_{aspect} = \{2, 4, 8\}$ Circular Contraction Flow

Once that prediction of experimental pressure-drops and their corresponding vortex-structure from Boger [1] and Boger et al. [2], have been achieved using the swIM model for an $\alpha_{aspect} = 8$, a parametric study is worthy to explore the predictive capabilities of this swanINNFM(q) family of models, in terms of the elusive task of predicting the appearance and persistence of lip-vortex structures in less demanding $\alpha_{aspect} = \{2, 4\}$ circular contraction geometries. Here, one may note that the versatile swIM model, with its control of normal-stress response, both in elongational (extensional-viscosity) and shear first normal-stress in shear deformations, is able to answer such a question. Specifically, different flow-structures and their evolution are observed by varying parameters of solvent-fraction β, extensibility L-parameter, and extensional dissipation-parameter λ_D. Each of these material parameters may be associated with corresponding variations in rheological properties (see Figure 3), observed through extensional viscosity η_e and normal-stress difference in shear N_{1Shear}. Under alternative parameter-variation, one notes the selection of implied base-values of: $L = 5$ for β–variation, and $\beta = 0.9$ for L-variation. Then, the dissipation-parameter may be set either at $\lambda_D = 0$ or $\lambda_D = 0.075$. From this perspective, counterpart findings may then be explored comparatively for flows in lower and alternative geometric contraction-ratios. Here, parameters influencing normal-stress response in both shear and extensional deformations appear with a key-role in the development of varied flow-structure formation in the contraction flows analyzed.

3.2.1. 8:1 Contraction Flow: Flow-Rate and Solvent-Fraction Adjustment ($1/9 \leq \beta \leq 0.9$)

This analysis starts with the $\alpha_{aspect} = 8$ circular contraction case, with focus on three flow-rate regimes that highlight essential vortex-structure features.

Low flow-rates. Wi = [1, 2] Under swIM [$L = 5$, $\lambda_D = 0.075$], only salient-corner vortex (*scv*) activity is recorded at $Wi = 1$ (Figure 7a), with no apparent lip-vortex (*lv*) activity. Then whilst remaining at $Wi = 1$, vortex-size and intensity strengthen considerably with solvent-fraction β-reduction. With increase in flow-rate level, at $Wi = 2$ (Figure 7b), the first appearance for *lv*-formation is found at $\beta = 0.7$, attendant with *scv*-presence. At lower solvent-fractions, such lip-vortex presence considerably enhances, to even take over the salient-corner vortex-intensity and produce a single strong elastic-corner vortex with a prominently convex separation-line; see Boger [1] and Boger et al. [2] for analogous experimental trends. Consistently, one observes *ecv* Ψ_{min}-intensity nearly doubling from $\beta = 0.5$ to $\beta = 0.3$; and tripling from $\beta = 0.5$ to $\beta = 1/9$, as recorded in Figure 7.

Medium Wi = [3, 5] and high [Wi = 16] flow-rates. Addressing the medium flow-rate regime, and comparing against that at low flow-rates, lip-vortex formation appears earlier with solvent-fraction β-decline. So, for example, at $Wi = 3$, first lip-vortex detection is noted with more dilute systems at $\beta = 0.9$, see Figure 8a. Flow-patterns and trends outlined at lower flow-rates are then repeated under solvent-fraction β-decrease. This precedes pronouncedly concave *ecv*-formation by $\beta \leq 0.6$. This theme is continued into the second yet higher-rate level in the medium flow-rate regime of $Wi = 5$, shown in Figure 8b. In the high flow-rate regime, still larger *ecv* Ψ_{min}-intensities abound in Figure 9, through some 8-times increase in vortex-size from $\beta = 0.9$ to $\beta = 1/9$ and dramatic vortex separation-line adjustment.

Extensional dissipative-parameter λ_D-variation. In passing, it is worth reflecting on the influence of dissipation extensional-influence and λ_D-rise on the results above; specifically in the medium flow-rate range $Wi = 3$, adopting other parameters of $L = 5$, $\beta = 0.9$. In Figure 10, this is illustrated through streamline-patterns across the range $0.1 \leq \lambda_D \leq 1.0$, in a regime of strong lip-vortex activity, both in intensity and spatial occupation. This is useful for insight upon the $\alpha_{aspect} = 4$ ratio case reviewed below. In this data, one can detect a clear amplification of the lip-vortex in the range $0.1 \leq \lambda_D \leq 0.4$, prior to coalescence with the salient-corner vortex ($0.5 \leq \lambda_D$), the lip-vortex being the dominant feature both prior to and post-coalescence. As a consequence, the eye of the vortex-center, subsequent to coalescence, is driven towards the re-entrant corner (response observed experimentally with Boger

Fluids 1 and 2; Boger [1]), announcing the onset of a phase of strong elastic corner-vortex domination, and further, unsteady vortex-oscillation at still larger flow-rates [8].

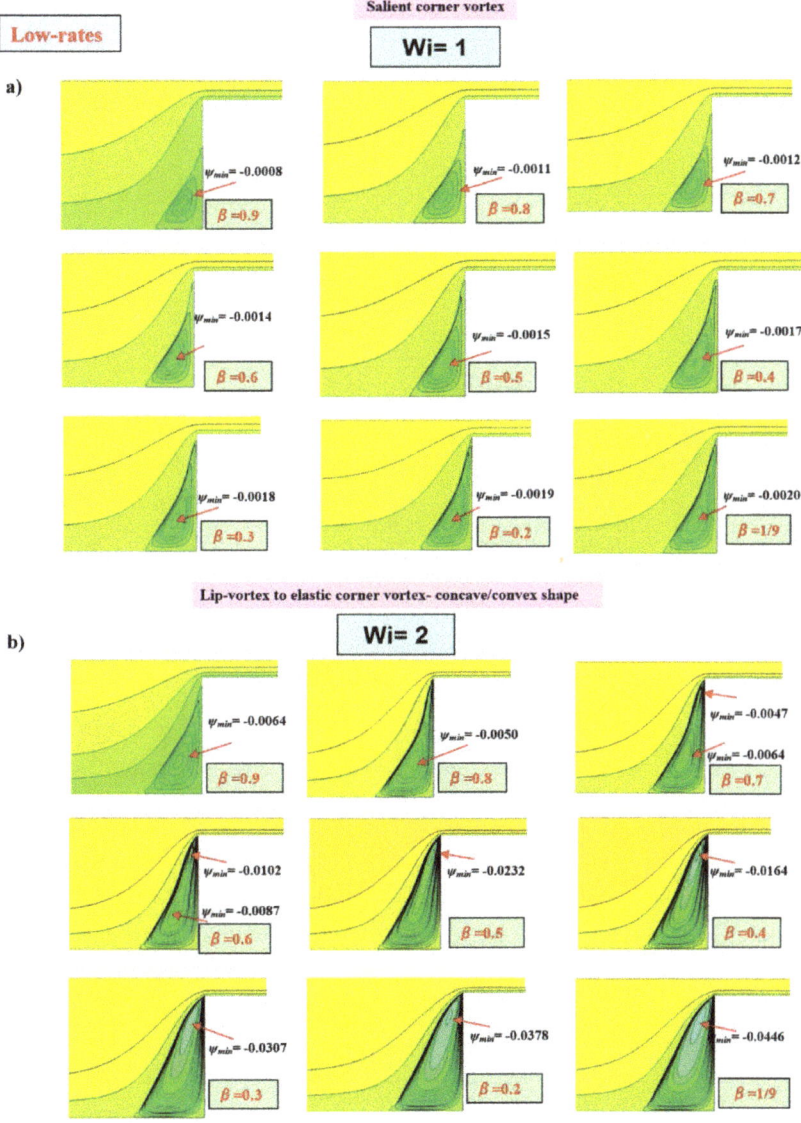

Figure 7. Streamlines; α_{aspect} = 8; low flow-rates: (**a**) Wi = 1, (**b**) Wi = 2; swIM [L = 5, λ_D = 0.075]; $1/9 \leq \beta \leq 0.9$.

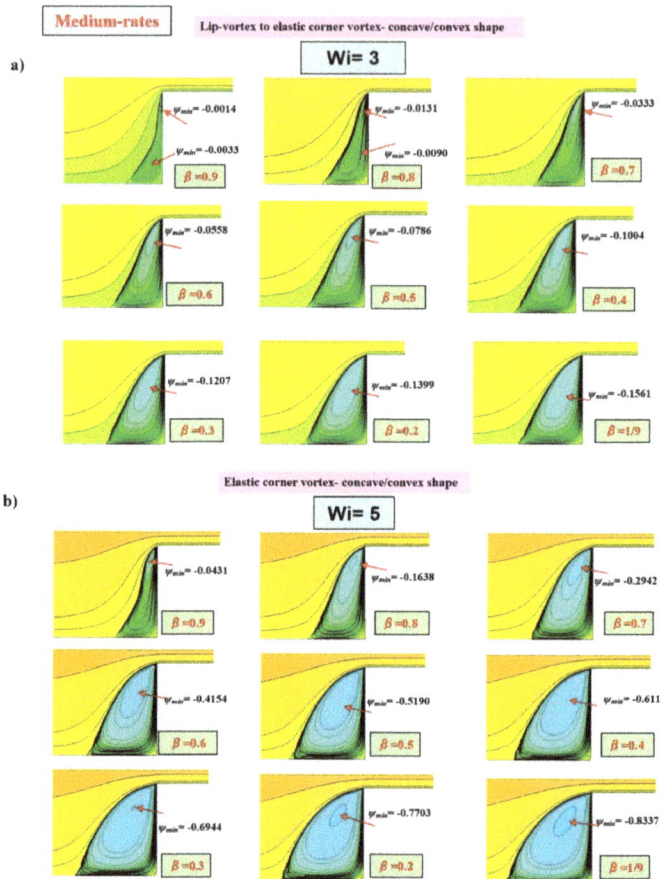

Figure 8. Streamlines; $\alpha_{aspect} = 8$; medium flow-rates: (**a**) $Wi = 3$, (**b**) $Wi = 5$; swIM [$L = 5$, $\lambda_D = 0.075$]; $1/9 \leq \beta \leq 0.9$.

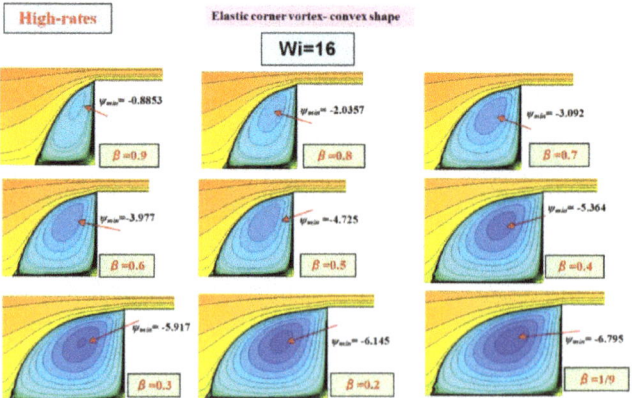

Figure 9. Streamlines; $\alpha_{aspect} = 8$; high flow-rates: $Wi = 16$; swIM [$L = 5$, $\lambda_D = 0.075$]; $1/9 \leq \beta \leq 0.9$.

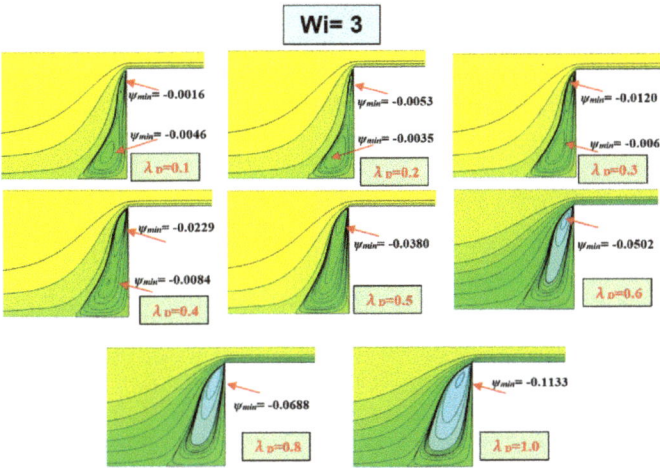

Figure 10. Streamlines; $\alpha_{aspect} = 8$; $Wi = 3$; swIM [$L = 5$, $\beta \leq 0.9$]; λ_D -rise.

3.2.2. Comparison across Geometric Aspect-Ratios α_{aspect}

Predictive solutions with swIM [$L = 5$, $\lambda_D = 0.1$, $\beta = 0.9$] provide direct comparison on major findings across the various contraction-ratios, taken against increasing flow-rate (Q) and charted through rise in Wi, according to low, moderate, and high flow-rate regimes. Note that to truly amplify the detail, scaling is applied in the cross-stream direction. Accordingly, streamline field-patterns are reported in Figure 11 and the line-graph of Figure 12. This data conveys the corresponding trend in movement on vortex-intensity. Notably, lip-vortex appearance is only recorded in the larger contraction-ratio $\alpha_{aspect} = 8$.

Low flow-rate ($Wi = 1$). In the first row of Figure 11, comparable vortex-structures are discerned across all three aspect ratios, gathering common concave-shaped separation-lines (referenced to the salient-corner recess), and with practically identical *scv* Ψ_{min}-intensity of $O(10^{-4})$.

Intermediate flow-rate regime [$Wi = 3$, $Wi = 5$]. Here, more interesting distinction can be drawn. At ($Wi = 3$), the second row of Figure 11, both $\alpha_{aspect} = \{2, 4\}$ solutions retain *scv* Ψ_{min}-intensity of $O(4 \times 10^{-3})$, yet proving one order-of-magnitude larger rotational intensity than at corresponding $Wi = 1$. With ($\alpha_{aspect} = 8$, $Wi = 3$), there is a relatively marginal decline in *scv*-intensity to $O(3 \times 10^{-3})$ noted, whilst also supporting some energy transfer into the onset of a lip-vortex, of intensity one order-of-magnitude lower, i.e., $O(5 \times 10^{-4})$. As above, here vortex separation-lines retain their concave-shaped form. At the more dynamic level of $Wi = 5$, corresponding to the third row of Figure 11, for $\alpha_{aspect} = 8$, there is a sudden burst of activity, with an increase of two orders-of-magnitude in lip-vortex rotational intensity, from $Wi = 3$ (row-two) to $Wi = 5$ (row-three). Simultaneously, the standing *scv*-intensity triples to $O(9 \times 10^{-3})$. Now for the first time, conspicuously, the vortex separation-line begins to adjust in shape around the lip-vortex zone, depicting somewhat of a more convex-to-concave delineation. In contrast, neither of the lower ratio $\alpha_{aspect} = \{2, 4\}$ solutions pick up any sign of lip-vortex activity, whilst their *scv*-intensities reflect levels comparable to those of {lip-vortex, $\alpha_{aspect} = 8$}, being slightly larger at $O(2 \times 10^{-2})$.

High flow-rate regime of $Wi = [10, 20]$. Rows four and five of Figure 11 now display vortex separation-lines of convex shapes for all three ratio-solutions. Nevertheless, the $\alpha_{aspect} = 8$ $Wi = 10$-solution (row-four of Figure 11) is disparate, in that its elastic-corner vortex has an evolution history that passes through coexistent *lv-scv* structures. In contrast, both $\alpha_{aspect} = \{2, 4\}$ solutions provide an *ecv* delivered from a growing salient-corner vortex directly, without any intermediate transition. The relative position of vortex-centre loci across contraction-ratio, clearly displays dependency upon their evolution history through flow-rate rise, with $\alpha_{aspect} = 8$ locating the *ecv* closer to the re-entrant corner. These vortex-evolution patterns concur well with Boger Fluid-1 (PAA/CS) findings of Boger [1]. Particularly notable is the trend observed for aspect-ratios ($\alpha_{aspect} \geq 4$), and the movement of the vortex-eye gravitating towards the re-entrant corner with increasing flow-rate (Figure 11).

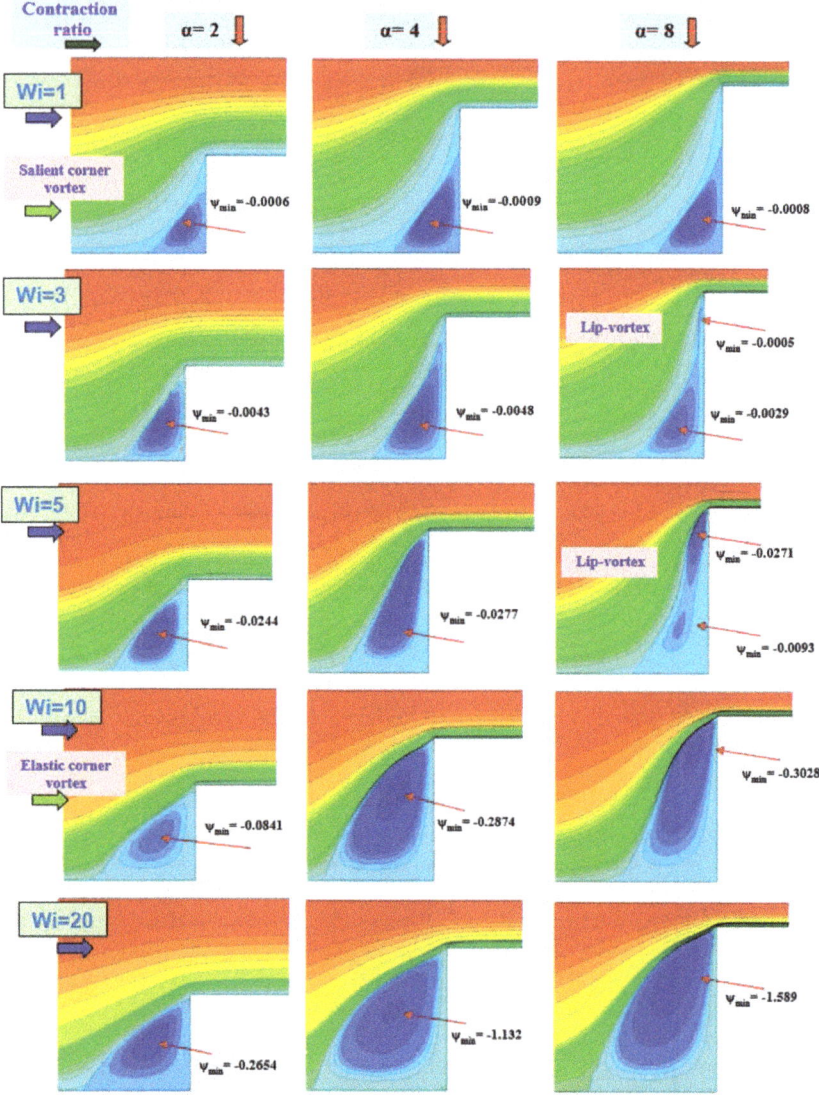

Figure 11. Streamlines; $\alpha_{aspect} = \{2, 4, 8\}$; $Wi = [1, 20]$; swIM [$L = 5, \beta \leq 0.9$]; $\lambda_D = 0.1$.

Informative overall trends can be gathered from vortex-intensity ($-\Psi_{min}$) data listed in Table 2 and its graphical representation in Figure 12. From a united trend at low flow-rates, there is a pronounced rise in vortex-intensity at ($Wi \geq 5$), in instances $\alpha_{aspect} = \{4, 8\}$ above $\alpha_{aspect} = 2$. By $Wi = 20$, separation is clearly apparent between all three instances, with the largest *ecv*-intensity attracted by $\alpha_{aspect} = 8$ (with *lv*-formation marked); and trends in $\alpha_{aspect} = 4$ solutions follow closely those under $\alpha_{aspect} = 8$.

Notably, trends in $\alpha_{aspect} = 2$ ratio solutions are the least dynamic in adjustment, providing smooth and continuous rise in *scv*-intensity.

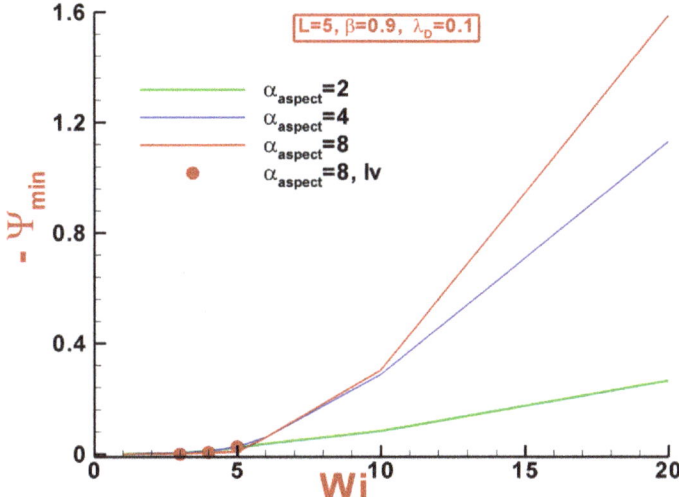

Figure 12. Vortex-intensity $(-\Psi_{min})$ against Wi; $\alpha_{aspect} = \{2, 4, 8\}$; swIM $[L = 5, \beta \leq 0.9]$; $\lambda_D = 0.1$.

Table 2. Vortex-intensity $(-\Psi_{min})$; $\alpha_{aspect} = \{2, 4, 8\}$; $Wi = [1, 20]$; swIM $[L = 5, \beta \leq 0.9]$; $\lambda_D = 0.1$.

Wi/α_{aspect}	2	4	8
1	0.0006	0.0009	0.0008
3	0.0043	0.0048	0.0005 0.0029(*lv*)
5	0.0244	0.0277	0.0093 0.0271(*lv*)
10	0.0841	0.2874	0.3028
20	0.2654	1.132	1.589

3.2.3. $\alpha_{aspect} = 4$ and $\alpha_{aspect} = 2$ Ratios: Lip-Vortices, Rise in Wi, Extensibility-Parameter L, and Solvent-Fraction β Switch

In this section, a parametric study on swIM extensibility parameter L and solvent fraction β is carried out, to discern the possibility of lip-vortex formation in less stringent contraction-ratios of $\alpha_{aspect} = \{2, 4\}$. Recall that Boger [1] and Boger et al. [2] only observed a lip-vortex experimentally for the PIB/PB-based fluid at $\{Wi^{exp} = \lambda^{exp}\dot{\gamma} = 2.3;\ Wi_{\lambda_1}^{comp} = 6.71\ Wi_{\lambda}^{exp} = 15.4\}$ under the $\alpha_{aspect} = 4.08$ ratio flow. This is performed through manipulation of extensional properties modulated by L-variation and solute-content promoted by β-decrease. Particularly, one turns to predictions for fluids with more pronounced extensional features and slightly higher solvent-fractions, i.e., $L = 12$, $\lambda_D = 0.1$, and $\beta = 0.7$ (as noted in Boger Fluids 1 and 2 of Boger [1]). One should mention that relatively more diluted fluids ($\beta = 0.8$; Figure 13) give signs of lip-vortex formation, but such kinematical structures remain difficult to track under such conditions.

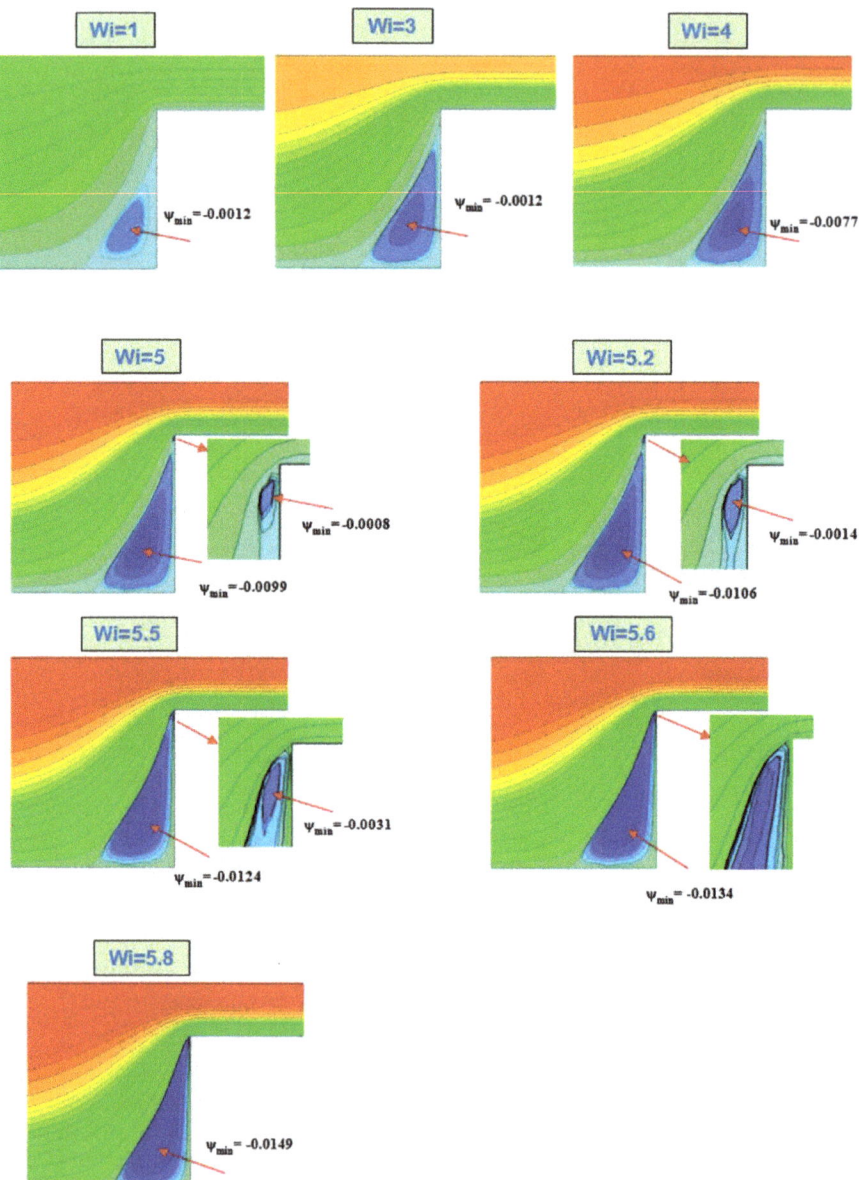

Figure 13. Vortex-intensity ($-\Psi_{min}$); $\alpha_{aspect} = 4$; swIM [$L = 12$, $\beta = 0.8$, $\lambda_D = 0.1$].

Lip-vortex-capture under $\alpha_{aspect} = 4$, $\beta = 0.7$, $L = 12$ and $\lambda_D = 0.1$ settings. Figure 14 illustrates a successful lip-vortex prediction under $\alpha_{aspect} = 4$. Firstly, an order-of-magnitude increase is detected in *lv*-intensity from $Wi = 4$ ($\psi_{min} = -0.0005$) to $Wi = 5.7$ ($\psi_{min} = -0.0053$). This proves to be the largest lip-vortex observed for the various different $\alpha_{aspect} = 4$ trial-settings. As such, $Wi = 5.7$ represents a critical level, beyond which steady-state solution-tractability fails. What is apparent is the fine balance in rheology here, between both solvent-fraction settings and hardening-levels. This interplay clearly has strong impact upon such localized issues as lip-vortex appearance (or not, as the case may be). Clearly experimentally, both such outcomes were observed in Boger [1]; with Boger Fluid-2 substantiating a bulb-like lip-vortex.

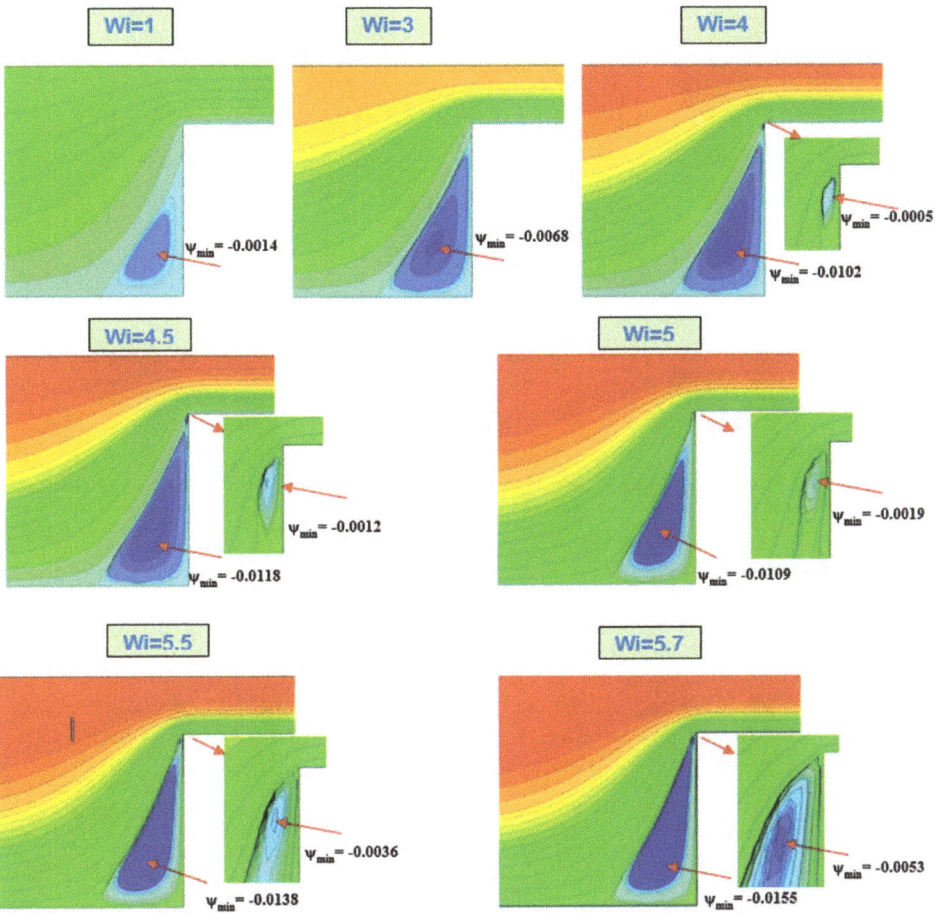

Figure 14. Vortex-intensity ($-\Psi_{min}$); $\alpha_{aspect} = 4$; swIM [$L = 12$, $\beta = 0.7$, $\lambda_D = 0.1$].

Seeking lip-vortices through λ_D-rise under $\beta = \{0.8, 0.7\}$ and $L = 12$. Following on from the parameter adjustment and findings on vortex activity in Tamaddon-Jahromi et al. [4] under $\alpha_{aspect} = 8$ flow, one may be peaked to further investigate the distinct influence of the extensional dissipation-parameter λ_D. In particular, seeking the segregated impact of extensional viscosity alone on $\alpha_{aspect} = 4$ lip-vortex response, as identified above. Figure 15 provides further evidence supplied with λ_D-rise, at the associated two solvent-fraction levels of $\beta = \{0.8, 0.7\}$. From this data, it is clear that early dissipation-factor λ_D-rise (with its η_e-boosting control) does strengthen lip-vortex activity, prior to this being subsumed by the more dominant salient-corner vortex, as the latter fingers its way towards the re-entrant corner. This response is simply intensified with reduction in solvent-fraction, with $\beta = 0.7$ subsuming lv-activity at $\lambda_D = 0.4$, whilst $\beta = 0.8$ performs likewise by $\lambda_D = 0.3$.

Fluids **2020**, *5*, 85

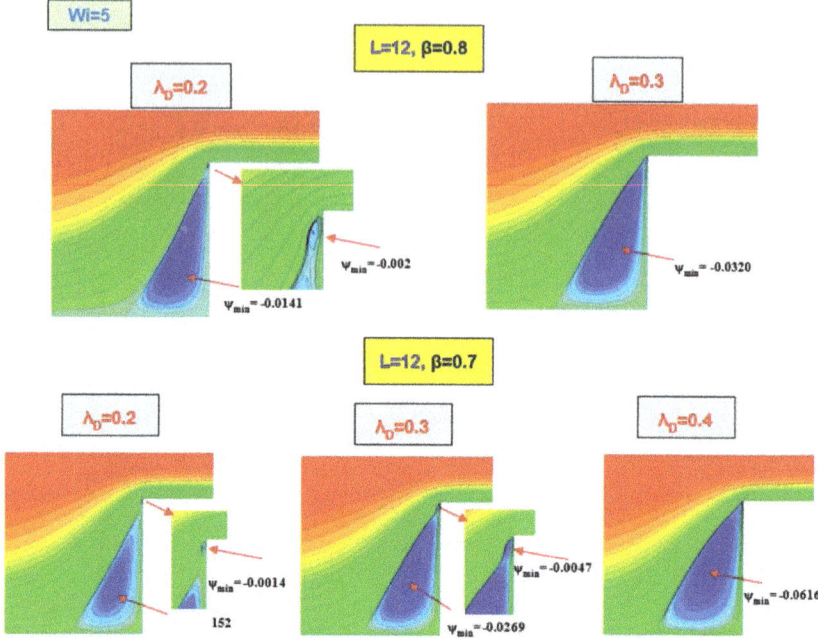

Figure 15. Vortex-intensity ($-\Psi_{min}$); α_{aspect} = 4; swIM [L = 12, β = {0.7, 0.8}, λ_D = {0.2, 0.3, 0.4}].

On α_{aspect} = 2 flow. Finally, the flow dynamics for the α_{aspect} = 2 contraction-ratio are so mild, that one needs to locate large polymeric-composition as high as β = 1/9 to detect any indication of *lv*-activity. Even then, such activity is limited to the temporal evolution phase alone, and vanishes prior to a steady-state being established (also noted by others for planar α_{aspect} = 4 contraction flow, see Sato and Richardson [17], Olsson [18]). In this instance, the influence of extensibility structure *L*-parameter adjustment on vortex-structures is barely noticeable.

4. Conclusions

Findings in this study go some way to replicating the experimental observations of Boger [1] and Boger et al. [2], on flow structures generated in three circular contraction flows of different aspect-ratios of α_{aspect} = {2, 4, 8}, and with two different Boger fluids, of essentially the same rheological response. Predictions reveal the delicate rheological balances at play, to match such response, taking into account the separate and combined influences of extensional viscosity and first normal-stress difference. Accordingly, elusive lip-vortices have been captured, as well as transition phases between salient and lip-vortices, their coalescence and the domination of elastic-corner vortices, and their relationship pressure-drops and first normal-stress difference in complex combined shear-to-extensional flow. Predictions permit fine interrogation of such phenomena. The flow history through flow-rate *Q*-rise provides the key to recognizing the dominant rheological characteristics here, where the swIM-model's precise tuning of normal-stress response on both shear and extension permits the capture of lip-vortices in the three geometrical configurations explored. Moreover, a link between flow-structure (through vortex-morphology and first normal-stress difference in the vicinities of the constriction) and pressure-drop enhancement staging is reported. In this respect, vortex-shape concurs well with N_1-field structure in the recess zones, for which connection between kinematics and rheological response is exposed. The pressure-drop enhancement mechanism, associated with the non-linear viscoelastic features of the Boger solutions, correlates well with vortex-dynamics phasing, where transition is recorded from early concave salient-corner vortices, coexistent lip and salient-corner vortices, and evolution to convex elastic-corner vortices, as reported experimentally [28] and numerically [9].

Of particular note across contraction-ratio $\alpha_{aspect} = \{2, 4, 8\}$ solutions is the gradually shifting upwards of the vortex-centre loci. It is attracted to the re-entrant corner, yet the relative positions of vortex-centre loci clearly display dependencies upon their salient-corner to elastic-corner vortex evolution history through flow-rate rise. These vortex evolution patterns concur well with the Boger Fluid-1 (PAA/CS) and Fluid-2 (PIB/PB) findings of Boger [1], prominent for aspect-ratios ($\alpha_{aspect} \geq 4$), and the movement of the vortex-eye gravitating towards the re-entrant corner with increasing flow-rate, finally leading to elastic-corner vortex domination and viscoelastic pressure-drop enhancement.

Author Contributions: Conceptualization, H.R.T.-J. and J.E.L.-A.; methodology, H.R.T.-J. and J.E.L.-A.; software, J.E.L.-A.; validation, H.R.T.-J. and J.E.L.-A.; formal analysis, H.R.T.-J. and J.E.L.-A.; investigation, H.R.T.-J. and J.E.L.-A.; resources, H.R.T.-J. and J.E.L.-A.; data curation, J.E.L.-A.; writing—original draft preparation, H.R.T.-J. and J.E.L.-A.; writing—review and editing, H.R.T.-J. and J.E.L.-A.; project administration, J.E.L.-A.; funding acquisition, J.E.L.-A. All authors have read and agreed to the published version of the manuscript.

Funding: This research was funded by Universidad Nacional Autónoma de México, Programa de Apoyo a Proyectos de Investigación e Innovación Tecnológica—Program of Support to Research Projects and Technological Innovation (UNAM-PAPIIT), grant number PAPIIT IA105620; by Facultad de Química, UNAM, Programa de Apoyo a la Investigación y Posgrado—Program of Support to Research and Post-graduate Studies, Facultad de Química, UNAM (UNAM-PAIP), grant number PAIP 5000-9172; and by Laboratorio Nacional de Cómputo de Alto Desempeño (LANCAD), Dirección General de Cómputo y de Tecnologías de la Información y Comunicación (DGTIC), UNAM, grant number LANCAD-UNAM-DGTIC-388, for the computational time provided on the *Miztli* Supercomputer.

Acknowledgments: Sincere thanks must be expressed for the many helpful contributions made to this study through our colleagues in the INNFM Wales, but particularly to Ken Walters FRS and Peter Townsend. Support from Consejo Nacional de Ciencia y Tecnología (CONACYT, Mexico) and from Perumal Nithiarasu, Deputy Head of College of Engineering and Director of Research, Swansea University, UK, are gratefully acknowledged.

Conflicts of Interest: The authors declare no conflict of interest.

References

1. Boger, D.V. Viscoelastic flows through contractions. *Ann. Rev. Fluid Mech.* **1987**, *19*, 157–182. [CrossRef]
2. Boger, D.V.; Hur, D.U.; Binnington, R.J. Further observations of elastic effects in tubular entry flows. *J. Non-Newton. Fluid Mech.* **1986**, *20*, 31–49. [CrossRef]
3. López-Aguilar, J.E.; Webster, M.F.; Tamaddon-Jahromi, H.R.; Walters, K. Numerical vs experimental pressure drops for Boger fluids in sharp-corner contraction flow. *Phys. Fluids* **2016**, *28*, 103104. [CrossRef]
4. Tamaddon-Jahromi, H.R.; López-Aguilar, J.E.; Webster, M.F. On modelling viscoelastic flow through abrupt circular 8:1 contractions—Matching experimental pressure-drops and vortex structures. *J. Non-Newton. Fluid Mech.* **2018**, *251*, 28–42. [CrossRef]
5. Nigen, S.; Walters, K. Viscoelastic contraction flows: Comparison of axisymmetric and planar configurations. *J. Non-Newton. Fluid Mech.* **2002**, *102*, 343–359. [CrossRef]
6. Binding, D.M.; Couch, M.A.; Walters, K. The pressure dependence of the shear and elongational properties of polymer melts. *J. Non-Newton. Fluid Mech.* **1998**, *79*, 137–155. [CrossRef]
7. Pérez-Camacho, M.; López-Aguilar, J.E.; Calderas, F.; Manero, O.; Webster, M.F. Pressure-drop and kinematics of viscoelastic flow through an axisymmetric contraction—Expansion geometry with various contraction-ratios. *J. Non-Newton. Fluid Mech.* **2015**, *222*, 260–271. [CrossRef]
8. Tamaddon-Jahromi, H.R.; Garduño, I.E.; López-Aguilar, J.E.; Webster, M.F. Predicting Excess pressure drop (epd) for Boger fluids in expansion-contraction flow. *J. Non-Newton. Fluid Mech.* **2016**, *230*, 43–67. [CrossRef]
9. López-Aguilar, J.E.; Webster, M.F.; Tamaddon-Jahromi, H.R.; Pérez-Camacho, M.; Manero, O. Contraction-ratio variation and prediction of large experimental pressure-drops in sharp-corner circular contraction-expansions—Boger fluids. *J. Non-Newton. Fluid Mech.* **2016**, *237*, 39–53. [CrossRef]
10. López-Aguilar, J.E.; Webster, M.F.; Tamaddon-Jahromi, H.R.; Manero, O.; Binding, D.M.; Walters, K. On the use of continuous spectrum and discrete-mode differential models to predict contraction-flow pressure drops for Boger fluids. *Phys. Fluids* **2017**, *29*, 121613. [CrossRef]

11. Webster, M.F.; Tamaddon-Jahromi, H.R.; López-Aguilar, J.E.; Binding, D.M. Enhanced pressure drop, planar contraction flows and continuous-spectrum models. *J. Non-Newton. Fluid Mech.* **2019**, *273*, 104184. [CrossRef]
12. Walters, K.; Webster, M.F. The distinctive CFD challenges of computational rheology. *Int. J. Numer. Meth. Fluids* **2003**, *43*, 577–596. [CrossRef]
13. White, S.A.; Gotsis, A.D.; Baird, D.G. Review of the entry flow problem: Experimental and numerical. *J. Non-Newton. Fluid Mech.* **1987**, *24*, 121–160. [CrossRef]
14. Owens, R.G.; Phillips, T.N. *Computational Rheology*, 1st ed.; Imperial College Press: London, UK, 2002.
15. Boger, D.V.; Binnington, R.J. Experimental removal of the re-entrant corner singularity in tubular entry flows. *J. Rheol.* **1994**, *38*, 333–349. [CrossRef]
16. Rothstein, J.P.; McKinley, G.H. The axisymmetric contraction-expansion: The role of extensional rheology on vortex growth dynamics and the enhanced pressure drop. *J. Non-Newton. Fluid Mech.* **2001**, *98*, 33–63. [CrossRef]
17. Sato, T.; Richardson, S.M. Explicit numerical simulation of time dependent viscoelastic flow problems by a finite element/finite volume method. *J. Non-Newton. Fluid Mech.* **1994**, *51*, 249–275. [CrossRef]
18. Olsson, F. A solver for time-dependent viscoelastic fluid flows. *J. Non-Newton. Fluid Mech.* **1994**, *51*, 309–340. [CrossRef]
19. Wapperom, P.; Webster, M.F. A second-order hybrid finite-element/volume method for viscoelastic flows. *J. Non-Newton. Fluid Mech.* **1998**, *79*, 405–431. [CrossRef]
20. Webster, M.F.; Tamaddon-Jahromi, H.R.; Aboubacar, M. Time-Dependent Algorithms for Viscoelastic Flow: Finite Element/Volume Schemes. *Numer. Meth. Par. Diff. Eq.* **2005**, *21*, 272–296. [CrossRef]
21. Aboubacar, M.; Webster, M.F. A cell-vertex finite volume/element method on triangles for abrupt contraction viscoelastic flows. *J. Non-Newton. Fluid Mech.* **2001**, *98*, 83–106. [CrossRef]
22. López-Aguilar, J.E.; Webster, M.F.; Tamaddon-Jahromi, H.R.; Manero, O. High-Weissenberg predictions for micellar fluids in contraction–expansion flows. *J. Non-Newton. Fluid Mech.* **2015**, *222*, 190–208. [CrossRef]
23. López-Aguilar, J.E.; Webster, M.F.; Tamaddon-Jahromi, H.R.; Manero, O. Convoluted models & high-Weissenberg predictions for micellar thixotropic fluids in contraction-expansion flows. *J. Non-Newton. Fluid Mech.* **2016**, *232*, 55–66. [CrossRef]
24. Chilcott, M.D.; Rallison, J.M. Creeping flow of dilute polymer solutions past cylinders and spheres. *J. Non-Newton. Fluid Mech.* **1988**, *29*, 381–432. [CrossRef]
25. White, J.L.; Metzner, A.B. Development of constitutive equations for polymeric melts and solutions. *J. Appl. Polym. Sci.* **1963**, *7*, 1867–1889. [CrossRef]
26. Debbaut, B.; Crochet, M.J. Extensional Effects in Complex Flows. *J. Non-Newton. Fluid Mech.* **1988**, *30*, 169–184. [CrossRef]
27. Debbaut, B.; Crochet, M.J.; Barnes, H.A.; Walters, K. Extensional effects in inelastic liquids. In Proceedings of the Xth International Congress on Rheology, Sydney, Australia, 14–18 August 1988; Australian Society of Rheology: Sidney, Australia, 1988; pp. 291–293.
28. Binding, D.M.; Walters, K. On the use of flow through a contraction in estimating the extensional the extensional viscosity of mobile polymer solutions. *J. Non-Newton. Fluid Mech.* **1988**, *30*, 233–250. [CrossRef]

© 2020 by the authors. Licensee MDPI, Basel, Switzerland. This article is an open access article distributed under the terms and conditions of the Creative Commons Attribution (CC BY) license (http://creativecommons.org/licenses/by/4.0/).

Article

Verification and Validation of *openInjMoldSim*, an Open-Source Solver to Model the Filling Stage of Thermoplastic Injection Molding

João Pedro, Bruno Ramôa, João Miguel Nóbrega * and Célio Fernandes

Institute for Polymers and Composites, University of Minho, Campus de Azurém, 4800-058 Guimarães, Portugal; b8389@dep.uminho.pt (J.P.); bruno.ramoa@dep.uminho.pt (B.R.); cbpf@dep.uminho.pt (C.F.)
* Correspondence: mnobrega@dep.uminho.pt

Received: 3 May 2020; Accepted: 25 May 2020; Published: 29 May 2020

Abstract: In the present study, the simulation of the three-dimensional (3D) non-isothermal, non-Newtonian fluid flow of polymer melts is investigated. In particular, the filling stage of thermoplastic injection molding is numerically studied with a solver implemented in the open-source computational library *OpenFOAM*®. The numerical method is based on a compressible two-phase flow model, developed following a cell-centered unstructured finite volume discretization scheme, combined with a volume-of-fluid (VOF) technique for the interface capturing. Additionally, the Cross-WLF (Williams–Landel–Ferry) model is used to characterize the rheological behavior of the polymer melts, and the modified Tait equation is used as the equation of state. To verify the numerical implementation, the code predictions are first compared with analytical solutions, for a Newtonian fluid flowing through a cylindrical channel. Subsequently, the melt filling process of a non-Newtonian fluid (Cross-WLF) in a rectangular cavity with a cylindrical insert and in a tensile test specimen are studied. The predicted melt flow front interface and fields (pressure, velocity, and temperature) contours are found to be in good agreement with the reference solutions, obtained with the proprietary software *Moldex3D*®. Additionally, the computational effort, measured by the elapsed wall-time of the simulations, is analyzed for both the open-source and proprietary software, and both are found to be similar for the same level of accuracy, when the parallelization capabilities of *OpenFOAM*® are employed.

Keywords: injection molding; filling stage; Cross-WLF model; Tait model; finite volume method; *openInjMoldSim*; *OpenFOAM*®

---

1. Introduction

The ever-growing demand for plastic materials promoted an increase in the requirements of many transformation processes, as happens for injection molding (IM), which is currently the most used one to manufacture thermoplastic parts. IM allows producing parts with a wide range of geometries and presents a return on investment in the medium/long term. Moreover, this transformation process presents a high fixed cost, not only due to the molds used, but also due to the many process variables that are necessary to control and optimize, to achieve the optimal processing conditions. It is exactly due to these requirements that simulation started to gain more importance in IM, being currently an indispensable step in the design phase to, among others, minimize the amount of experimental iterations needed to manufacture a tool, to decrease the overall cost of development, and to be able to produce parts with the desired geometrical and mechanical requirements.

There are several proprietary software solutions available for the simulation of the IM process and related techniques [1]. However, in addition to the expensive licenses, the proprietary software usually cannot be adapted to specific user needs, thus preventing the access of small to medium sized companies to these tools. Open-source free alternatives may allow turning around these limitations, and, additionally, one can have access and adapt the source code. In the broad area of computational fluid dynamics (CFD), *OpenFOAM*® [2] became a very interesting alternative to proprietary software, due to the range of available continuum mechanics solvers and an active development community.

In what concerns to polymer processing, *OpenFOAM*® [2] has been used in different areas like extrusion, blow molding, and to some extent, injection molding. In the extrusion area, many *OpenFOAM*® [2] based tools are already available; for example, the work in Habla et al. [3] presented the development and validation of a model to compute the temperature distribution in the calibration stage of extruded profiles. Later, the work in Rajkumar et al. [4] presented design guidelines to support die designers to obtain balanced flow distributions. For the numerical modeling of IM, Araújo et al. [5] provided a parallel 3D unstructured non-isothermal flow solver, with both polymer and air being considered as incompressible fluids. Their results presented a good accuracy and reasonable parallel efficiency and scalability. Making use of the *OpenFOAM*® [2] computational library, the work in Nagy and Steinbichler [6] introduced a framework of a compressible two-phase (air and polymer melt) fluid model, with polymer-specific material models, for the description of the filling, packing, and cooling stages. Although for simple geometries, they proved that the framework of solving the compressible form of continuity, linear momentum, and energy equations resulted in a good agreement between experimental observations and theoretical predictions. Other researchers also tried to succeed in this task. Magalhães [7] in her Master Thesis implemented the energy equation, as well as the Bird–Carreau rheological model in the *interFoam* solver available in the *OpenFOAM*® [2] CFD library. She studied the filling stage of the IM process and implemented a boundary condition, which allowed the air to escape from the cavity, but not the polymer, as happens in real-life molds. With the use of that boundary condition in the numerical simulations, she concluded that a higher percentage of filled cavity was obtained. She also showed that differences in polymer and air viscosities were some of the main reasons for the large number of instabilities observed in the numerical calculation procedure. Recently, also aiming to model the IM process, the works in Mole et al. [8] and Krebelj and Turk [9] implemented both the Cross-WLF and modified Tait models in the *compressibleInterFoam* (a solver designated by *openInjMoldSim*). The objective was to simulate the filling and packing stages, considering compressible flow, and applying appropriate models for the polymeric materials' behavior. The solver was validated with a simple 2D demo test case for an amorphous polystyrene grade. Subsequently, Fontaínhas [10] in her Master Thesis used the *openInjMoldSim* solver with a 3D case study (a simple thin rectangular geometry) and compared the results predicted by the solver (velocity, pressure, and temperature profiles) with the ones obtained by the proprietary software *Moldex3D*® [11]. A mesh sensitivity analysis was performed, and she reached the conclusion that the results predicted with the *openInjMoldSim* solver were qualitatively similar to the ones obtained with *Moldex3D*® [11]. However, the results obtained with the mesh refinement study did not converged to a level of accuracy with statistical significance. She also reported that the open-source solver was 18 times slower than the proprietary one.

This work describes the numerical validation of an open-source solver, *openInjMoldSim* [8,9], able to simulate the filling stage of the thermoplastic IM process. For that purpose, mesh sensitivity studies were performed for all the validation case studies. This way, accurate/converged solutions were obtained for all the flow fields, which could not be found in preliminary studies performed with the same numerical algorithm. For that purpose, we used the Richardson extrapolation technique to estimate the discretization error. Additionally, to the best of the authors' knowledge, this was the first time that compressible and incompressible formulations were compared, and the effect of the artificial interface compression term was assessed for the IM filling stage. Finally, to conclude about the novel contributions of the present work, the studies performed provided a detailed comparison of open-source and proprietary software, in terms of accuracy and performance.

This paper is organized in the following manner: In Section 2, we present the mathematical formulation of the modeling code that is used in this work, including the governing equations and numerical discretization. In Section 3, we present a number of case studies involving the filling of cavities with Newtonian and generalized Newtonian fluids. In each case study, a comparison of the results obtained with *openInjMoldSim* is performed with data reported in the literature and obtained with the proprietary software *Moldex3D*® [11], in order to assess its accuracy, robustness, and performance. Finally, in Section 4, we summarize the main conclusions of this work.

2. Mathematical Formulation

In this section, we present the mathematical formulation for an algorithm that is able to handle the polymer melt filling stage of the IM process efficiently. Contrarily to the commonly used assumption of fluids' incompressibility [12], the employed algorithm considers a compressible, non-isothermal based formulation, which, in the future, will provide a rigorous basis for the simulation of the packing and cooling stages. The implementation was performed in the open-source framework code *OpenFOAM*® [2,9].

2.1. Governing Equations

In the approach employed, the air and the polymer melt phases are assumed to be immiscible and compressible, unless otherwise stated. During the filling stage, the flow is governed by the mass, momentum, and energy conservation equations, which can be written as follows [1]:

$$\frac{\partial \rho}{\partial t} + \nabla \cdot (\rho \mathbf{u}) = 0, \tag{1}$$

$$\frac{\partial (\rho \mathbf{u})}{\partial t} + \nabla \cdot (\rho \mathbf{u}\mathbf{u}) = \nabla \cdot \sigma, \tag{2}$$

$$\frac{\partial (\rho c_P T)}{\partial t} + \nabla \cdot (\rho c_P \mathbf{u} T) = \beta T \left(\frac{\partial p}{\partial t} + \mathbf{u} \cdot \nabla p \right) + p \nabla \cdot \mathbf{u} + \sigma : \nabla \mathbf{u} + \nabla \cdot (k \nabla T), \tag{3}$$

where $\mathbf{u}$, p, and T denote the velocity vector, pressure, and temperature fields, respectively, t is the time, and σ is the total, or Cauchy, stress tensor. The parameters ρ, β, c_P, and k represent the fluid density, compressibility coefficient, specific heat capacity, and thermal conductivity, respectively. The Cauchy stress tensor is given by:

$$\sigma = -p\mathbf{I} + \tau, \tag{4}$$

where $\mathbf{I}$ is the identity tensor and τ is the extra or deviatoric stress tensor, which, for the flow of a compressible generalized Newtonian fluid, is defined as:

$$\tau = 2\eta (T, p, \dot{\gamma}) \mathbf{D} - \frac{2}{3} \eta (T, p, \dot{\gamma}) (\nabla \cdot \mathbf{u}) \mathbf{I}, \tag{5}$$

where $\dot{\gamma} = \sqrt{2(\mathbf{D} : \mathbf{D})} = \sqrt{2 \, \text{tr}(\mathbf{D}^2)}$ is the generalized shear rate, with $\mathbf{D}$ being the deformation rate tensor defined as $\mathbf{D} = (\nabla \mathbf{u} + (\nabla \mathbf{u})^T)/2$, and $\eta (T, p, \dot{\gamma})$ is the fluid viscosity, which is considered to be a function of T, p and $\dot{\gamma}$ (see Section 2.1.1 for detailed information about the rheological models used in this work).

The simulation of the IM filling stage requires solving the fluid flow equations and capturing the melt front location at every time step. For that purpose, an advanced free-surface capturing model [13], based on the volume-of-fluid (VOF) method [14] is used. In this method, a species transport equation is used to track the relative volume fraction of the two phases, or phase fraction, α, distribution. The phase fraction ranges from zero to one, zero being for air and one for the polymer melt. In this way, the fluid interface is located in regions where $0 < \alpha < 1$, and the physical properties can be calculated as an average weighted by α, as follows:

$$\psi = \psi_m \alpha + \psi_a (1 - \alpha), \tag{6}$$

where the subscripts m and a denote the melt and air phases' properties, respectively, and ψ represents all the relevant fluid properties, namely ρ, η, c_P, and k. The governing equation for the phase fraction, α, is defined as:

$$\frac{\partial \alpha}{\partial t} + \nabla \cdot (\alpha \mathbf{u}) + \nabla \cdot [\alpha(1-\alpha)\mathbf{u}_r] = S_p + S_u, \tag{7}$$

where $\mathbf{u}_r$ is the relative velocity vector, commonly denominated by "compression velocity". In this work, the compression velocity is given by:

$$\mathbf{u}_r = \mathbf{n}_f \, min \left[C_\alpha \frac{|\phi|}{|\mathbf{S}_f|}, max \left(\frac{|\phi|}{|\mathbf{S}_f|} \right) \right], \tag{8}$$

where ϕ is the volume flux at faces, $\mathbf{S}_f$ is the face normal vector pointing outwards from the cell, and C_α is an adjustable coefficient that determines the magnitude of the compression (usually $C_\alpha \geqslant 1$) [15]. $\mathbf{n}_f$ is the interface unit normal vector taken from the phase fraction distribution as:

$$\mathbf{n}_f = \frac{(\nabla \alpha)_f}{\left|(\nabla \alpha)_f + \delta\right|}, \tag{9}$$

where δ is a stabilization factor [15].

The artificial interface compression term is an extra term added to the phase fraction evolution equation and is non-zero only when $0 < \alpha < 1$, i.e., at the fluid interface, and avoids its numerical diffusion (notice that in reality, the two fluid flow presents a sharp interface) [16]. Otherwise, a significant diffusion of α at the interface region might occur, predicting a nonphysical smeared interface, especially with coarse meshes. S_p and S_u represent two terms that arise from considering a compressible material [16]. Detailed information about the advanced free-surface capturing model used in this work can be found in Berberović et al. [13].

2.1.1. Constitutive Models

The rheological behavior of the melt and air phases is defined by a single-fluid two-phase model, where the average kinematic viscosity is given by Equation (6), in which ψ is replaced by the viscosity η. For the air phase, the kinematic viscosity is assumed to be constant, i.e., a Newtonian fluid. For the polymer melt, the Cross-WLF constitutive model is employed. This constitutive model is widely adopted for studying both the filling and packing stages of the IM process [17] and is given by:

$$\eta_m (T, p, \dot{\gamma}) = \frac{\eta_0 (T, p)}{1 + (\eta_0 (T, p) \dot{\gamma}/\tau^*)^{1-n}}, \tag{10}$$

where τ^* is the material constant representing the shear stress, above which the pseudoplastic behavior of the melt is observed, n is the power-law index, and $\eta_0 (T, p)$ is the viscosity at the null shear rate. In this work, we adopted the Williams–Landel–Ferry (WLF) model [18] to describe the dependence of η_0 on T and p, given by:

$$\eta_0 (T, p) = D_1 \exp \left[\frac{-C_1 (T - T_0)}{C_2 + T - T_0} \right], \tag{11}$$

where $T_0 = D_2 + D_3 p$ and D_1, D_2, D_3, C_1, and C_2 are parameters determined by experimental characterization [6].

2.1.2. Equation of State

Polymers shrink when pressure increases; thus, their specific volume (the reciprocal of the density) decreases [19]. On the other hand, with the increase of temperature, polymeric materials tend to expand, and, consequently, their specific volume increases. The relation of specific volume, pressure, and temperature is designated by the pressure-volume-temperature behavior (PVT) [20]. The most commonly used model to characterize the material PVT behavior of polymer melts is the modified Tait model [21], because it can describe properly the PVT relation of both amorphous and semi-crystalline polymers [19]. In the modified Tait model, the specific volume, $\widehat{V}$, is given by:

$$\widehat{V} = \widehat{V_0}\left[1 - C\ln\left(1 + \left(\frac{p}{B}\right)\right)\right] + \widehat{V_t}, \tag{12}$$

where:

$$\widehat{V_0} = \begin{cases} b_{1S} + b_{2S}(T - b_5), & \text{if } T \leq T_t \\ b_{1L} + b_{2L}(T - b_5), & \text{if } T > T_t \end{cases}, \tag{13}$$

$$B = \begin{cases} b_{3S}\exp(-b_{4S}(T - b_5)), & \text{if } T \leq T_t \\ b_{3L}\exp(-b_{4L}(T - b_5)), & \text{if } T > T_t \end{cases}, \tag{14}$$

$$\widehat{V_t} = \begin{cases} b_7\exp(b_8(T - b_5) - b_9 p), & \text{if } T \leq T_t \\ 0, & \text{if } T > T_t \end{cases}, \tag{15}$$

where $C = 0.0894$ is a dimensionless constant, the coefficients b_1 to b_9 are obtained by fitting experimental characterization data, and p and T are the material pressure and temperature, respectively [19]. The subscripts S, L, and t represent solid, liquid (or melt), and transition states, respectively. The transition temperatures for semi-crystalline and amorphous polymers are the melt (T_m) and glass (T_g) transition temperatures, respectively.

The equation of state employed for the air phase is the ideal gas equation given by:

$$\rho = \frac{p}{RT}, \tag{16}$$

where R is the ideal gas constant ($R = 8.314$ J/(K.mol)).

2.2. Numerical Discretization

The numerical solution of the equations presented in Section 2.1 relies on a collocated finite volume method (FVM) approach, which uses the conservative integral forms of the governing equations for the single-fluid two-phase model [22]. The implementation of the three conservation equations (continuity, momentum, and energy) is done using the *OpenFOAM*® computational library [2,23]. For more details about the solver employed in this work, the reader is referred to Mole et al. [8] and Krebelj and Turk [9].

3. Validation Case Studies

This section presents the case studies employed for the validation of the solver applied to the filling stage of the IM process (*openInjMoldSim*) presented in Section 2. The first case study was devoted to the flow along a cylindrical channel, in which the fluid was considered to have Newtonian rheology. The objective was to make a comparison of the obtained numerical velocity and pressure fields with an analytical solution available for steady-state conditions [24]. In the second case study, an assessment of the effect of the material compressibility in the filling stage of the IM process, in a rectangular cavity with a cylindrical insert, is presented. Finally, a detailed comparison between *openInjMoldSim* and *Moldex3D*® [11] was performed in a more complex geometry, the cavity of a mold employed to manufacture tensile test specimens, a part representative of the typical ones produced by the IM process, which was used for material mechanical characterization purposes [24,25].

In all case studies considered in this work, the domain was divided into inlet, outlet, and boundary walls. Moreover, a fixed velocity was imposed at the inlet boundary patch, and at the same location, the homogeneous Neumann boundary condition was applied for the pressure field. Additionally, the no-slip velocity was imposed at the cavity walls. Furthermore, the walls only experience heat flux modeled by heat convection. Since the initial and boundary conditions were similar for all case studies, for organization purposes, Table 1 shows the default initial and boundary conditions for all case studies. The particularities of each case study will be presented in the respective subsection.

Table 1. Initial and boundary conditions for all case studies.

	Temperature (T)	Velocity (U)	Pressure (P)	Phase Fraction (α)
Inlet	230 °C	U_{in}	Null normal gradient	1
Outlet	Null normal gradient	Null normal gradient	10^5 Pa (atmospheric)	Null normal gradient
Walls	Heat convection T_{mold} = 50 °C h = 1250 W/(m². K)	Null velocity	Null normal gradient	Null normal gradient
Initial Condition	230 °C	Null velocity	10^5 Pa (atmospheric)	0

The material used in the simulations was the GPPS (general purpose polystyrene) Styron 678, from Americas Styrenics. The rheological model employed was the Cross-WLF presented in Section 2, and its parameters are given in Table 2. These parameters were obtained from the *Moldex3D*® [11] materials database. The variation of viscosity with shear-rate, temperature, and pressure is given in Figure 1, which shows that the viscosity of the polymer melt decreases with the increase of both the shear-rate and the temperature.

Table 2. Cross-WLF coefficients for GPPS (general purpose polystyrene) Styron 678 from Americas Styrenics [11].

Parameter	Value
n	0.2903
τ^*(Pa)	13,678
D_1(Pa.s)	7.44×10^{10}
D_2(K)	373.15
D_3(K/Pa)	0.0
A_1	25.971
A_2(K)	51.6

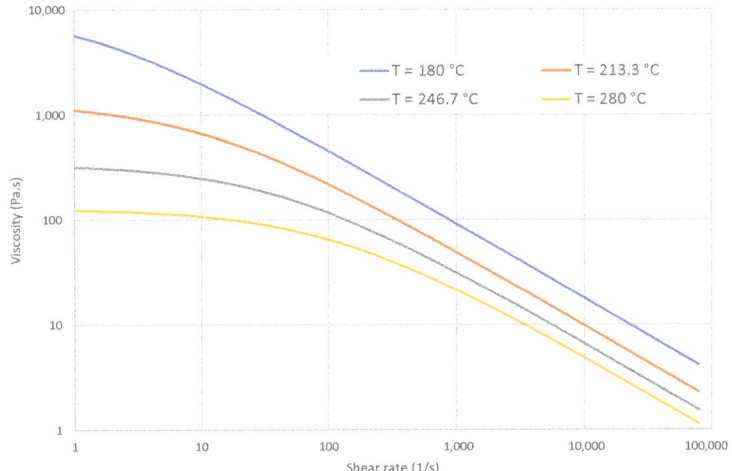

Figure 1. Variation of the viscosity with shear-rate and temperature for the GPPS Styron 678.

As stated before, the air was assumed to be a Newtonian fluid by making use of the Newtonian rheological model, and by setting a dynamic viscosity equal to 0.1 Pa.s. Additionally, the thermodynamic properties, specific heat capacity, c_p, and thermal conductivity, k, were defined as 1007 J/(kg.K) and 0.0263 W/(m.K), respectively, for the air and 2100 J/(kg.K) and 0.15 W/(m.K), respectively, for the polymer melt.

The PVT behavior of the material was modeled using the modified Tait model, illustrated in Figure 2. The compressibility model adopted show that the specific volume of the material decrease with the increase of pressure, and increase with the increase of temperature following a slope until T_g (glass transition temperature) is reached. After T_g, the slope increases (as is commonly observed in amorphous polymers).

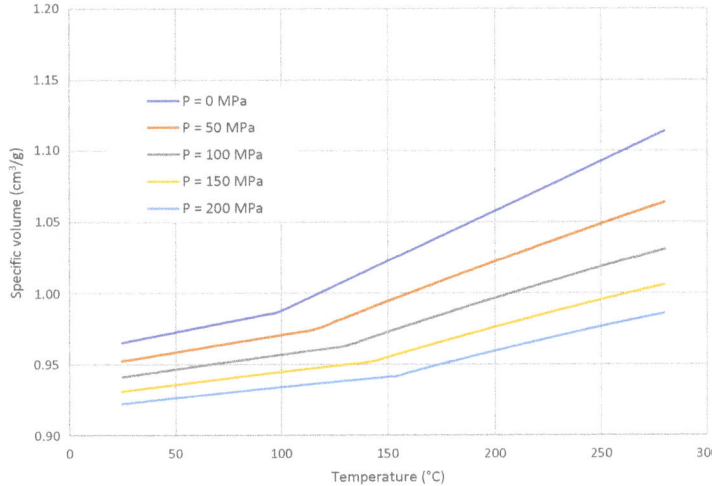

Figure 2. Variation of the polymer specific volume with temperature and pressure fields for the GPPS Styron 678.

The coefficients of the modified Tait model are shown in Table 3. These coefficients were obtained from the $Moldex3D^{®}$ materials database [11].

Table 3. Modified Tait model coefficients for the GPPS Styron 678 from Americas Styrenics [11].

Parameters	Compressible Formulation
b_{1m} (m³/kg)	0.0009881
b_{2m} (m³/(kg.K))	7.03×10^{-7}
b_{3m} (Pa)	1.71×10^{8}
b_{4m} (1/K)	0.004495
b_{1s} (m³/kg)	0.0009873
b_{2s} (m³/(kg.K))	2.89×10^{-7}
b_{3s} (Pa)	2.43×10^{8}
b_{4s} (1/K)	0.003106
b_{5} (K)	373.98
b_{6} (K/Pa)	2.88×10^{-7}
b_{7} (m³/kg)	0
b_{8} (1/K)	0
b_{9} (1/Pa)	0

In the analysis of the case studies presented below, the methodologies used are common to more than one case study. Therefore, for organization purposes, the methods for assessment and verification used are presented here:

IC and NIC variations: The abbreviation "IC" stands for interface compression, which means that the artificial interface compression term of Equation (7) is considered, and the abbreviation "NIC" stands for the simulation where it was neglected.

Switch-over time: The time for the polymer melt to reach the switch-over position. This is an important instant for the analysis of the numerical calculation, since it coincides with the transition of the filling to the packing/holding stages. Assuming incompressibility, the equation to calculate the time the polymer melt takes to reach the switch-over point, t_{SO}, is given by:

$$t_{SO} = \frac{V}{A \times U_{in}}, \quad (17)$$

where U_{in} is the imposed average velocity at the inlet face, A the area of the inlet face, and V the volume at the switch-over point. In all the case studies presented, the switch-over point was assumed to occur when 98% of the total cavity volume is filled, which is a value commonly employed in the IM process [26].

Average properties: For assessment purposes, throughout the three case studies, minimum, average, and maximum properties were calculated. Considering a number of cells n, the average values are given by:

$$\bar{\theta} = \sum_{i=1}^{n} \frac{(|C_i| \times \theta_i)}{|C_i|}, \quad (18)$$

where $|C_i|$ can be either the cell face area, when analyzing a cross-section, or the cell volume, when analyzing all the geometry. When $|C_i|$ is constant Equation (18) becomes the arithmetic mean.

Richardson's extrapolation (RE): This is a widely employed method to estimate discretization errors [27]. It can be applied to any discretization procedure, either for differential or integral governing equations, such as the FVM [28]. This method was used in this work to extrapolate the value of the variable under analysis to an infinitely refined mesh and, thus, allowed estimating the accuracy of the results at each level of mesh refinement. For a detailed description of this technique see Appendix A.

3.1. Case Study 1: Filling of a Cylindrical Cavity

This first case study covered the flow of a Newtonian fluid along a cylindrical channel and was used to verify if the numerical algorithm was properly implemented. The cavity geometry employed in this first case study is illustrated in Figure 3.

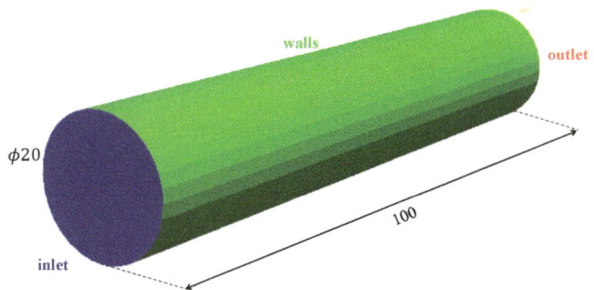

Figure 3. Geometry of the cylindrical cavity for Case Study 1 (dimensions in mm).

The initial and boundary conditions used in this case study are presented in Table 1, with an inlet velocity of $U_{in} = 4$ m/s.

The mesh sensitivity analysis was undertaken with three levels of refinement, where the number of cells of the coarsest mesh (M1) was doubled in each direction to obtain the second degree of refinement (M2). The same procedure was applied to obtain the most refined level (M3). The software *cfMesh* [29] was used to generate the three meshes, which had 4368, 34,280, and 263,016 cells for M1, M2, and M3, respectively.

The polymer melt was considered to behave as a Newtonian fluid, with a dynamic viscosity value of 310.0 Pa.s, obtained by using $n = 1$ in the Cross-WLF rheological model presented in Section 3. The thermodynamic properties of the air and polymer melt are also the ones presented in Section 3.

The PVT behavior of the material was prescribed using two formulations, the compressible one, given by the modified Tait model, which was already shown in Figure 2 and Table 3, and the incompressible, where the melt density employed was constant and equal to 1.075 g/cm³.

The first tests carried out were related to the time predicted by the algorithm to achieve the switch-over point, t_{SO}, for the incompressible formulation and to compare it with the theoretical value calculated from Equation (17). The results obtained are given in Table 4 for both the IC and NIC variations.

Table 4. Comparison of the switch-over time obtained for the incompressible formulation, considering both interface compression (IC) and neglecting IC (NIC) variations. The analytical reference value calculated from Equation (17) is $t_{SO} = 0.0245$ s.

Mesh	IC		NIC	
	t_{SO} (s)	Error (%)	t_{SO} (s)	Error (%)
M1	0.0249	1.630	0.0462	88.60
M2	0.0246	0.408	0.0250	2.040
M3	0.0230	6.120	0.0247	0.816

From the data shown in Table 4, one can conclude that the IC gave accurate results for coarse and medium meshes, because it minimizes the effect of the interface diffusion and, therefore, favors the mass conservation. Since NIC omits the contribution of the artificial interface compression, the results deviated from the analytical values, especially for the coarsest mesh (M1). On the other side, with refined meshes, neglecting the artificial diffusion term did not affect the mass balance. In a future work, a detailed study should allow identifying the criteria related to the necessity of considering the interface compression term in these simulations.

To understand the differences caused by considering the polymer melt as incompressible, and since the compressible formulation was the most realistic one, Table 5 shows the switch-over time obtained for both formulations, when different mesh refinement levels were employed and using the NIC variation, which gave the best results with the incompressible formulation (see Table 4). In this case, both formulations, compressible and incompressible, predicted similar results, with the former filling the cavity slightly faster than the latter on M2 and M3. This happened because the (volumetric) flow rate was imposed at the inlet, and with the compressible formulation, due to the pressure increase at that location, the actual mass flow rate was slightly higher. The small differences obtained between both formulations allowed concluding that the incompressible formulation gave accurate results in what concerns to the instant predicted for the end of the filling stage.

Table 5. Comparison of the switch-over time obtained with the compressible (reference) and incompressible formulations, with the NIC variation.

Mesh	t_{SO} (s)		Relative Difference (%)
	Compressible (Reference)	Incompressible	
M1	0.0462	0.0462	0.000
M2	0.0249	0.0250	0.402
M3	0.0246	0.0247	0.407

To further investigate the differences between the incompressible and compressible formulations, an analytical solution for the velocity profiles at steady-state conditions was employed. This solution was presented by Liang et al. [24] for the power-law model. Thus, considering the power-law exponent equal to one, the velocity profile for Newtonian fluids is given by:

$$u(r) = \frac{r_0}{2}\left(\frac{r_0 \Delta p}{2\mu L}\right)\left[1 - \left(\frac{r}{r_0}\right)^2\right] \quad (19)$$

where $\frac{\Delta p}{L}$ is taken as the pressure gradient throughout the channel length, which means that Δp is the difference between the pressure at the inlet and outlet boundary patches, L and r_0 are the channel length and radius, respectively, and μ is the fluid viscosity. r is the radius at a specific point where the velocity profile is taken. The inlet face average pressure was computed with Equation (18).

Table 6 shows the maximum velocity values obtained with Equation (19) and normalized by the inlet velocity, the extrapolated value obtained using Richardson's extrapolation technique (Equation (A1)), and the errors (Equation (A3)) at each level of mesh refinement employed, using both compressible and incompressible formulations.

Table 6. Maximum velocity values for the mesh refinement study for both compressible and incompressible formulations, Richardson's extrapolation (RE), and the relative errors.

Mesh	Compressible		Incompressible	
	Velocity/U_{in}	Error (%)	Velocity/U_{in}	Error (%)
M1	1.912	4.37	1.906	4.43
M2	1.973	1.33	1.968	1.35
M3	1.992	0.40	1.986	0.41
RE	2.000	—	1.995	—

These results show that the velocity values presented a monotonic and asymptotic behavior with mesh refinement for both formulations, and therefore, the errors for each level of mesh refinement decreased. Moreover, the velocity values were very close between both formulations; however, the ones obtained with the compressible formulation were always higher. This happened because to keep the volumetric flow rate constant at the inlet boundary, more material was introduced into the cavity, to compensate the shrinkage caused by the larger pressures exerted at that location.

Table 7 shows the pressure values taken at the center of the channel inlet boundary patch, for both incompressible and compressible formulations and all levels of mesh refinement, the Richardson's extrapolation value (Equation (A1)), and the relative errors (Equation (A3)).

Table 7. Pressure values at the center of the channel inlet boundary patch, for both incompressible and compressible formulations and all levels of mesh refinement, Richardson's extrapolation value (RE), and the relative errors.

Mesh	Compressible		Incompressible	
	Pressure (MPa)	Error (%)	Pressure (MPa)	Error (%)
M1	4.49	5.96	4.46	5.93
M2	4.65	2.61	4.61	2.77
M3	4.72	1.14	4.68	1.29
RE	4.77	—	4.74	—

As happened for the velocity values, the pressure values presented a monotonic and asymptotic behavior, with the errors for each level of mesh refinement decreasing towards the extrapolated value. Although the values of pressure were very close for both formulations, the compressible one presented always higher values, which agreed with the fact that in this formulation the cavity was filled faster than in the incompressible counterpart, which resulted in a higher pressure.

From the results presented, we concluded that considering an incompressible formulation did not significantly affect the accuracy of the final results at the end of the filling stage of the IM process, when considering a Newtonian constitutive model to describe the polymer rheological behavior.

3.2. Case Study 2: Filling of a Rectangular Cavity with a Cylindrical Insert

This case study covers the filling of a rectangular cavity with a cylindrical insert, illustrated in Figure 4. The cavity geometry has a constant thickness of 4 mm, a width of 40 mm, and a length of 150 mm, while the cylindrical insert has a diameter of 15 mm, and its center is located at 55 mm from the inlet. The boundary walls are also shown in Figure 4. In terms of flow behavior, this case study presents the separation and merge of the flow front, due to the cylindrical insert, forming a weld line [26]. Weld lines created by the merge of independent flow fronts are typical in IM [26].

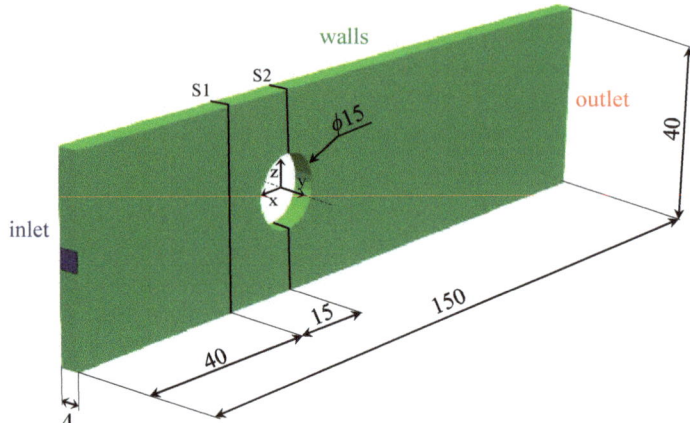

Figure 4. Geometry and boundary patches for Case Study 2 (dimensions in mm). S1 and S2 denotes Slice 1 and Slice 2, respectively.

The initial and boundary conditions for this second case study are presented in Table 1, with an inlet velocity of $U_{in} = 1.46$ m/s.

The mesh refinement strategy employed in this case study was similar to that of Case Study 1, meaning that three degrees of refinement were used, multiplying by a factor of two the number of cells in each direction, in consecutive mesh refinement levels. In this case study, meshes M1, M2, and M3 were generated using the *cfMesh* [29] utility, and comprise 23,712, 187,480, and 1,494,064 cells, corresponding to nearly 4, 8, and 16 cells along the cavity thickness, respectively.

For this case study, the polystyrene GPPS Styron 678 and the air properties employed in the numerical simulations were the ones defined in Section 3. The same compressible and incompressible formulations used in Case Study 1 were also simulated in this case study.

The results obtained using all the meshes considered did not present substantial differences; therefore, only the ones corresponding to the most refined mesh (M3) are presented. Velocity and temperature contours were taken from two cross-sections at different geometry locations (see Figure 4), representative of the initial flow channel region (Slice 1 (S1)) and at the cylindrical insert (Slice 2 (S2)). These results correspond to the switch-over point, i.e., the end of the filling stage. Figure 5 shows the velocity and temperature contours for S1 and S2 and both fluid formulations.

In S1, the location that was close to the cylindrical insert, the melted material at the center of the channel flowed at a lower velocity than the one that was closer to the top and bottom mold walls. This shows that the material flow division, motivated by the insert, already started in S1. In terms of temperature, as expected, the polymer melt at the center of the channel was hotter than the one at the walls (temperature profiles across the channel thickness are shown in Figure 6). Comparing both material formulations, there are no visible differences, both on the velocity and temperature distributions. The results obtained for S2, the location at the middle of the cylindrical insert, show a symmetric velocity profile because of the channel symmetry. As expected, the velocity magnitude increased because the material had a smaller cross-section to flow through. In what concerns to the temperature field, since the cross-section area was smaller, the maximum temperature slightly exceeded 230 °C (see also Table 8), which is the temperature imposed at the inlet, a direct effect of the viscous dissipation. Moreover, again, the two formulations show no visible differences, for both velocity and temperature distributions.

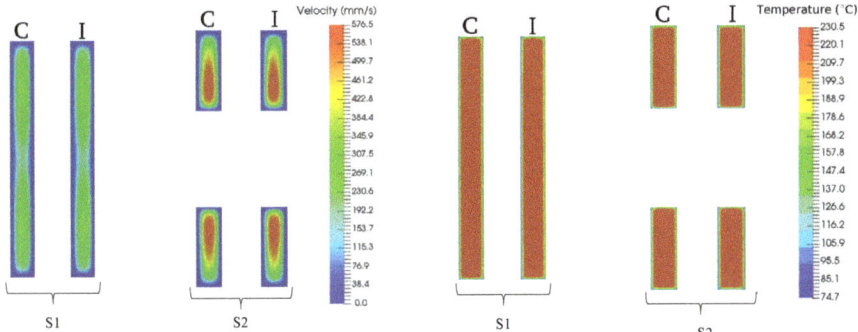

Figure 5. Velocity magnitude (left) and temperature (right) contours on slices S1 and S2 for Case Study 2. C, compressible formulation; I, incompressible formulation.

Figure 6 shows the temperature profile across the channel thickness at slices S1 (on $z = 0$ mm) and S2 (on $z = 10$ mm), for the incompressible fluid formulation. The results obtained with the compressible formulation were similar to the ones presented in Figure 6 and, therefore, are not shown to avoid the clustering of results. Due to the quasi-parabolic velocity profile that is obtained across the thickness, the region of maximum shear-rate, or viscous dissipation, occurs at the wall ($y = \pm 2$ mm). However, due to the boundary conditions considered for the temperature field, near the wall, the heat is removed from the polymer melt. As an outcome of this framework, the maximum temperature is obtained at a small distance from the wall. These facts are more evident in the inset provided on Figure 6, which also shows that due to the higher viscous dissipation, motivated by the steeper velocity profile, the maximum temperature was slightly higher in S2 than in S1, in accordance with the maximum temperature values provided in Table 8 (230.4 °C in S1 and 230.6 °C in S2).

In order to further check the effect of the incompressible formulation, the minimum, average, and maximum values for both fields (temperature and velocity magnitude) were computed for both slices (S1 and S2) at the switch-over point, and the results are presented in Tables 8 and 9.

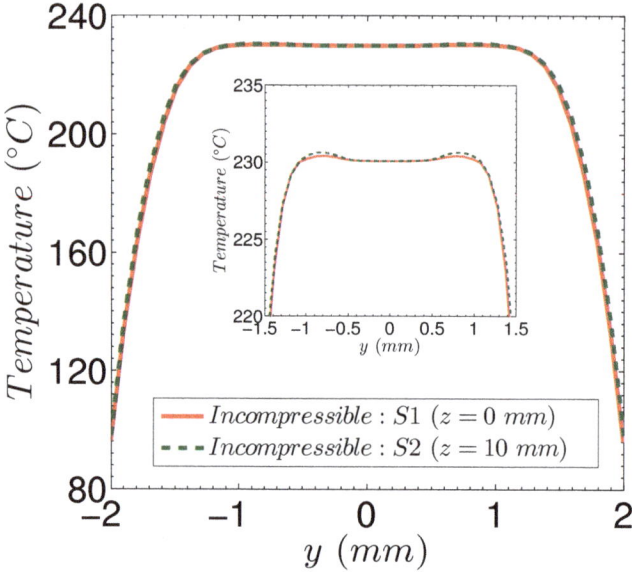

Figure 6. Temperature profiles across the channel thickness at slices S1 and S2, using the incompressible formulation, for Case Study 2.

Table 8. Minimum, average, and maximum temperature values for slices S1 and S2 (see Figure 4), at the switch-over point.

Slice	Formulation	Temperature (°C)			Relative Differences (%)		
		T_{min}	T_{ave}	T_{max}	T_{min}	T_{ave}	T_{max}
S1	C (ref)	98.0	214.4	230.2	—	—	—
	I	94.4	214.1	230.4	3.7	1.4	0.1
S2	C (ref)	103.3	213.5	230.3	—	—	—
	I	100.0	213.3	230.6	3.2	0.09	0.1

Table 9. Average and maximum velocity magnitude values for slices S1 and S2 (see Figure 4), at the switch-over point.

Slice	Formulation	Velocity (mm/s)		Relative Differences (%)	
		U_{ave}	U_{max}	U_{ave}	U_{max}
S1	C (ref)	148	342	—	—
	I	149	345	0.7	0.9
S2	C (ref)	233	570	—	—
	I	233	576	0.0	1.0

Notice that the minimum value of the velocity field magnitude was not presented in Table 9 because on the mold walls, the no-slip boundary condition was applied for the velocity field. From the results shown in Tables 8 and 9, we concluded that the differences between the compressible and incompressible formulations in both velocity and temperature fields are negligible, i.e., smaller than 4%.

Figure 7 shows the distribution of the pressure field on the mold cavity, computed for both compressible and incompressible formulations, at the switch-over point. These results showed that the maximum pressure required to fill the cavity, predicted by both formulations, was quite similar, with a difference of approximately 0.9%.

From all the results presented, we could conclude that both incompressible and compressible formulations were adequate to be used in the simulation of the filling stage of the IM process, with the former being more advantageous in terms of computational cost. The wall time spent by the incompressible formulation to reach the switch-over point was 39 h, while for the compressible formulation it was 41 h, circa 5% higher. In terms of stability, both formulations, compressible and incompressible, presented similar and good stability during the calculations, due to the use of the PIMPLE (combination of the two acronyms PISO and SIMPLE, while PISO stands for Pressure–Implicit withSplitting of Operators and SIMPLE stands for Semi–Implicit Method for Pressure Linked Equations) algorithm, which allowed obtaining converged solutions in each time-step.

Figure 7. Distribution of the pressure field on the mold cavity, for the two formulations, compressible (C) and incompressible (I), at the switch-over point.

3.3. Case Study 3: Filling of a Tensile Test Specimen

This third case study comprises the simulation of the filling stage of a tensile test specimen. The aim of this case study is to compare the accuracy and performance of *openInjMoldSim* with the proprietary software *Moldex3D®* [11]. This case study geometry, illustrated in Figure 8, is representative of the industrial practice, and corresponds to a specimen used for material mechanical characterization purposes [24,25]. To enlarge the scope of the analysis, the feeding system was included in the simulated geometry. Polymer melt enters the cavity through the inlet boundary located at the top of the sprue. The melted material then flows through the main runner and, subsequently, enters in the secondary runner, which changes the flow melt direction. Before reaching the specimen cavity, the melt flows through a very thin and small channel called the gate [19]. Finally, the melt fills the actual mold cavity, which has a constant thickness of 4 mm, typical of an injection molded part. The geometry dimensions are given in Figure 8. The boundary patches employed in this case study were similar to the ones of the previous case studies and are also represented in Figure 8. There is an inlet face from which the polymer melt enters and an outlet face from which air exited, while the remaining faces were impermeable walls that only experienced heat flux between the mold and the polymer melt.

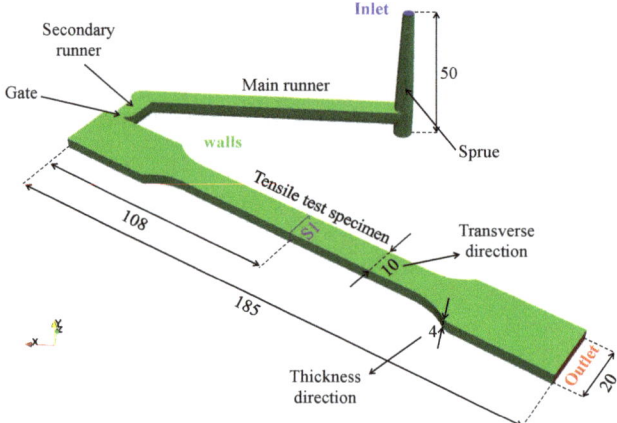

Figure 8. Geometry and boundary walls for Case Study 3 (dimensions in mm).

The initial and boundary conditions, prescribed in the $OpenFOAM^{®}$ [2] library, are presented in Table 1, with an inlet velocity of $U_{in} = 1.4$ m/s. In $Moldex3D^{®}$ [11], being proprietary software, many features cannot be defined by the user, as will be shown below for the pressure field. Anyway, the flow rate imposed in $Moldex3D^{®}$ [11] and $OpenFOAM^{®}$ [2] was the same and equal to $Q = 15.7$ cm^3/s. The temperature boundary conditions were also the same. The pressure field could not be accessed in the proprietary software and, therefore, was the only boundary condition that could not be imposed to be equal in both software.

Since the objective of this study was to make a comparison between the two software, it would be useful to perform the simulations with the same meshes. However, as $Moldex3D^{®}$ [11] only accepts meshes generated in its workbenches, and $OpenFOAM^{®}$ [2] could not import those meshes. Consequently, the simulations could not be made exactly in the same meshes. Therefore, the approach employed was to make simulations in meshes with a similar number of cells, until reaching mesh converged results. The same procedure employed in the previous case studies for the mesh sensitivity analysis was also followed here. Tables 10 and 11 describe the meshes used in $openInjMoldSim$ and $Moldex3D^{®}$ [11], respectively. The utility $cfMesh$ [29] was used to generate the meshes for $openInjMoldSim$; specifically, the Cartesian mesh application was employed to generate the majority of hexahedral cells. The $Moldex3D^{®}$ [11] meshes were created using the program workbench $Moldex3D\ R16\ Designer$ [11], applying a boundary layer mesh (BLM). The need to have an additional level of mesh refinement in $Moldex3D^{®}$ [11] will be explained below.

Table 10. $openInjMoldSim$ meshes employed in Case Study 3.

Mesh	Number of Cells	Cells along Transverse Direction	Cells along Thickness Direction
M1	30,091	8	5
M2	272,149	18	11
M3	2,110,987	37	21

Table 11. Moldex3D®[11] meshes employed in Case Study 3.

Mesh	Number of Cells	Cells along Transverse Direction	Cells along Thickness Direction
M1	29,625	8	2
M2	272,409	12	5
M3	2,101,139	26	10
M4	15,304,010	54	21

Tables 10 and 11 show that the number of cells for each mesh refinement level are very close for both software. Additionally, the number of cells for each direction of the tensile test specimen was duplicated in consecutive mesh refinement levels, and consequently, the total number of cells presented a factor of approximately eight. A detailed view of mesh M2 is illustrated in Figure 9 for both software.

Figure 9. Mesh M2 for *Moldex3D*® [11] (left) and *openInjMoldSim* (right).

The fluids employed in the simulation of the filling of the tensile test specimen were equal to the ones presented in the previous case study, and their properties are given in the introduction of Section 3. Again, the polymer melt was modeled as a generalized Newtonian fluid, while the air was assumed to present a Newtonian behavior.

The aim of the mesh sensitivity analysis was to identify the refinement level required to obtain mesh independent results. To verify the level of refinement needed to ensure the desired accuracy, the Richardson's extrapolation [30] was applied to the pressure field. Table 12 shows the values of the maximum pressure generated in the tensile test specimen cavity, the respective extrapolated value (RE), and the associated errors for both software. The values considered in this analysis were the ones corresponding to the switch-over point.

Table 12. Maximum pressure values obtained with the proprietary software *Moldex3D*® [11] and the open-source software *openInjMoldSim* at the switch-over point. The Richardson's extrapolated (RE) values and the relative errors are also presented.

Mesh	*Moldex3D*® [11]		*openInjMoldSim*	
	Maximum Pressure (MPa)	Error (%)	Maximum Pressure (MPa)	Error (%)
M1	7.18	43.4	7.76	29.6
M2	8.32	34.4	14.07	27.8
M3	9.66	23.8	12.01	9.1
M4	11.61	8.41	-	-
RE	12.68	-	11.01	-

From Table 12, it is possible to understand the reason for the existence of four levels of mesh refinement in *Moldex3D*® [11]. For this software, the error for mesh M3 was still very high (>20%); therefore, a fourth mesh was employed, which allowed reaching the same order of accuracy of *openInjMoldSim*. It is important to notice that the maximum values are more difficult to compute, since they tend to have larger errors, due to the fact that they were local values and, therefore, presenting larger sensitivity to numerical errors.

Figure 10 shows the contour of the pressure field distribution in the cavity at the switch-over point. The pressure profiles obtained in both software are qualitatively identical, when using the most refined meshes for both (M4 for *Moldex3D*® and M3 for *openInjMoldSim*). However, as also shown in Figure 10, the melt flow front predicted by the open-source software seemed more realistic than that of the proprietary counterpart, which present a plug-like surface.

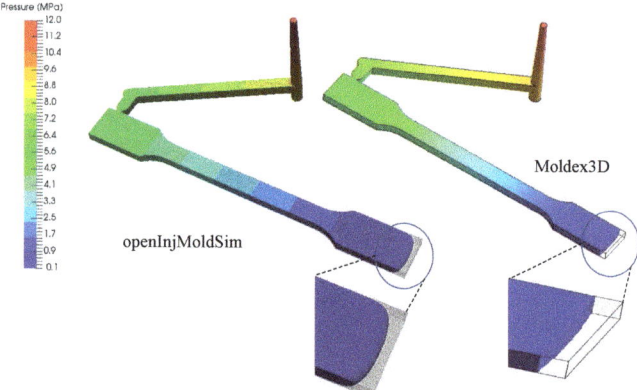

Figure 10. Pressure distribution at the switch-over point and melt flow front shape predicted by the most refined mesh of both software.

Additionally, as illustrated in Figure 11, the inlet pressure evolution along time was also monitored and compared between both software, using the results obtained with the most refined mesh.

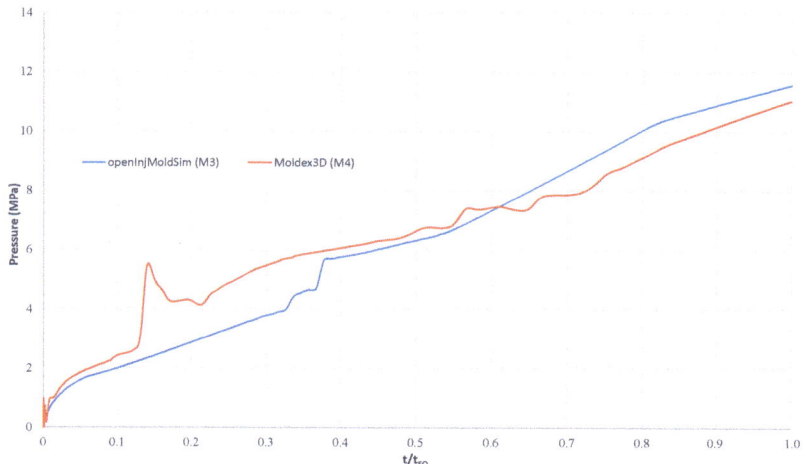

Figure 11. Time evolution of the inlet pressure obtained with the most refined mesh for both *openInjMoldSim* and *Moldex3D*® [11] software.

The melt pressure evolution presented some perturbations, which were compared with the progression of the melt flow front. The time intervals where the pressure evolution was perturbed were $0.1 < t/t_{SO} < 0.2$, $0.3 < t/t_{SO} < 0.4$, and $0.5 < t/t_{SO} < 0.8$. The evolution of the melt flow front for the first interval is illustrated in Figure 12. This perturbation in the pressure field was only predicted by *Moldex3D*® [11], but this behavior seems strange because at $t/t_{SO} = 0.1$, the melt flow front was already at the main runner.

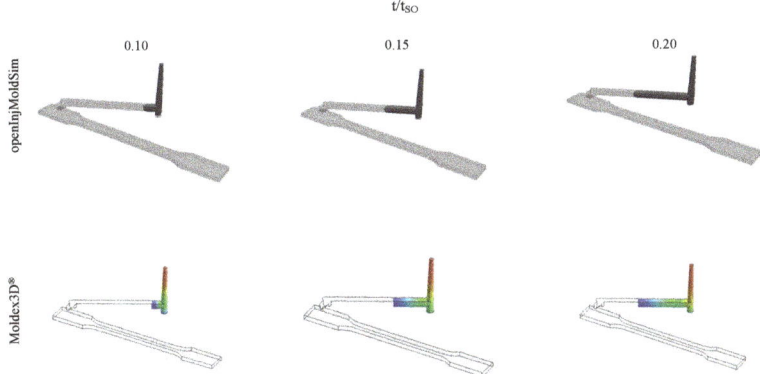

Figure 12. Melt front location for $t/t_{SO} = 0.1$, 0.15, and 0.2 predicted by both *openInjMoldSim* and *Moldex3D*® [11].

The melt flow front evolution for the second perturbation identified on the inlet pressure evolution is shown in Figure 13, and it coincides with the flow of the melt through the gate. This second perturbation was only predicted by the *openInjMoldSim* software.

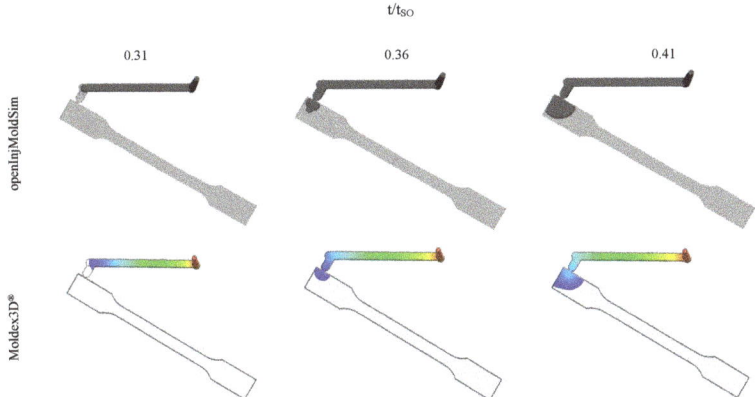

Figure 13. Melt front location for $t/t_{SO} = 0.31$, 0.36, and 0.41 predicted by both *openInjMoldSim* and *Moldex3D*® [11].

As explained before, the gate was a very small and thin channel; thus, it presented a high resistance to the flow, which justifies the change in the slope on the pressure evolution [19].

The flow front location for the last perturbation observed in the time evolution of the inlet pressure is illustrated in Figure 14. As shown, this interval corresponds to the flow of the melt front through the thinner region of the tensile test specimen. As the cross-section is narrower in that location, a change of the slope in the pressure evolution was expected.

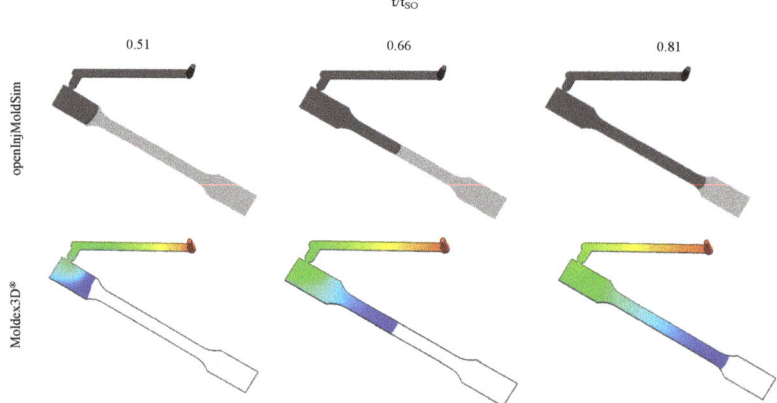

Figure 14. Melt front location for $t/t_{SO} = 0.51$, 0.66, and 0.81 predicted by both *openInjMoldSim* and *Moldex3D*® [11].

The graph presented in Figure 11 shows that for *openInjMoldSim*, the slope of the curve was higher in the third interval, when compared to the periods before and after it. In *Moldex3D*® [11] for M4, the pressure evolution curve presents several nonphysical oscillations; however, the pressure is still increasing, as expected.

Finally, a general overview of the accuracy and performance of both software can be obtained from the data presented in Table 13, where the errors that correspond to the three variables (pressure, velocity, and temperature), as well as the execution time for each level of mesh refinement are given.

Table 13. General comparison of the accuracy and performance of both *openInjMoldSim* and *Moldex3D*® [11].

Software	Mesh	Errors (%)			Number of Cores	Execution Time
		P_{max}	U_{max}	T_{max}		
Moldex3D® [11]	M1	43.4	14.2	0.7	1	88 s
	M2	34.4	19.6	0.6	1	23 min
	M3	23.8	20.4	0.5	8	3.8 h
	M4	8.41	7.22	0.2	8	24.5 h
openInjMoldSim	M1	29.6	33.2	1.0	1	12.8 h
	M2	27.8	26.7	0.1	8	41 h
	M3	9.1	8.2	0.0	48	98.5 h
					96	59 h
					192	34 h

From the results shown in Table 13, one can conclude that, in general, *openInjMoldSim* present better accuracy than *Moldex3D*® [11], with the exception of the temperature field, where the errors are similar. In order to reach the same accuracy for all fields, *Moldex3D*® [11] needed at least one more degree of mesh refinement than *openInjMoldSim*. It should be noticed that for the proprietary software, the results did not converge asymptotically with the mesh refinement, which forced to use the Richardson's extrapolation approach given by Equation (A1), and estimate the apparent order.

In terms of performance, the proprietary software was clearly faster than the open-source for the same level of refinement. *Moldex3D*® [11] was 523, 107, and 9 times faster than *openInjMoldSim*, respectively for M1, M2, and M3, although they were not computed with the same number of cores. However, when analyzing degrees of mesh refinement for the same accuracy, M3 for *openInjMoldSim* and M4 for *Moldex3D*® [11], the proprietary software was only 1.4 times faster than the open-source one; however, the number of cores used was not the same. Notice that, since *Moldex3D*® [11] is a proprietary software, there is a restricted number of licenses, which did not allow us to solve the problem with more cores. On the other side, since in *OpenFOAM*® [2] the only limitation is the computational resources available, if the number of cores are increased, a similar calculation time could be obtained.

4. Conclusions

A numerical formulation available in the open-source computational library *OpenFOAM*® [2], the *openInjMoldSim* solver, was assessed in different test cases related to the simulation of the filling stage of the injection molding process.

The verification/validation of the open-source solver was performed with three case studies, where the results obtained with the open-source solver were compared both with analytical solutions, for simpler case studies, and with results obtained with *Moldex3D*® [11], a proprietary software widely employed in the industry.

The first case study allowed identifying the more appropriate modeling setup and to understand some details of the numerical algorithm developed to describe the injection molding process, by considering the simplified Newtonian constitutive model. The results obtained with this first case study allowed us to conclude that the artificial interface compression term, presented in the volume of fluid formulation, provided better results for coarse and medium meshes, but in refined meshes, it affected the mass conservation negatively. Furthermore, the open-source solver predictions presented a good agreement with the analytical reference solutions. Finally, the compressible formulation did not present substantial differences in relation to the incompressible one.

In the second case study, two formulations for the equation of state (compressible and incompressible) based on the modified Tait model were used. This study allowed verifying that in the presence of incompressible formulations, the performance of the numerical algorithm was improved without a significant loss of accuracy. This observation further indicated that incompressible formulations were a good approximation for the filling stage of the injection molding process and may be a way of increasing the overall calculation efficiency.

In the third case study, the results obtained with *openInjMoldSim* and *Moldex3D*® [11] were compared. The open-source software presented results that practically converged, i.e., with relative errors below 10% within three levels of mesh refinement. Additionally, its accuracy was better than the proprietary software for each level of mesh refinement. In order to reach the same accuracy obtained with *openInjMoldSim*, *Moldex3D*® [11] required an additional level of mesh refinement. Moreover, the latter did not present asymptotic convergence of the errors with the mesh refinement. Furthermore, a few nonphysical results were obtained with the proprietary software, both in terms of the flow front shape, and in the evolution of inlet pressure, in which, for instance, the flow of the melt front through the gate could not be identified. The open-source software did not present similar limitations.

All these results led to the conclusion that the *openInjMoldSim* solver presented a better accuracy than *Moldex3D*® [11]. In terms of performance, the scenario was reversed, with *Moldex3D*® [11] being clearly faster than the *openInjMoldSim* solver. However, one of the main limitations of the proprietary software is the number of licenses available. Therefore, if there were enough computational resources available, open-source software could match the calculation speed of the proprietary one.

In summary, the results presented here showed that the injection molding code, developed using an open-source framework, could be used to model the filling stage of injection molding processes accurately. However, additional studies should be made to improve the calculation performance.

Author Contributions: Conceptualization, J.M.N. and C.F.; Methodology, J.P., B.R., J.M.N. and C.F.; Formal Analysis, J.P., J.M.N. and C.F.; Writing—Original Draft Preparation, J.P. and C.F.; Writing—Review & Editing, J.P., B.R., J.M.N. and C.F.; Supervision, J.M.N. and C.F.; Project Administration, J.M.N.; Funding Acquisition, J.M.N. All authors read and agreed to the published version of the manuscript.

Funding: This work is funded by FEDER funds through the COMPETE 2020 Programme and National Funds through FCT (Portuguese Foundation for Science and Technology) under the projects UID-B/05256/2020, UID-P/05256/2020, MOLDPRO-Aproximações multi-escala para moldação por injeção de materiais plásticos (POCI-01-0145-FEDER-016665), and FAMEST-Footwear, Advanced Materials, Equipment's and Software Technologies (POCI-01-0247-FEDER-024529).

Acknowledgments: The authors would like to acknowledge the Minho University cluster under the project NORTE-07-0162-FEDER-000086 and the Minho Advanced Computing Center (MACC) for providing HPC resources that contributed to the research results reported within this paper.

Conflicts of Interest: The authors declare no conflicts of interest.

Appendix A

Richardson's extrapolation is usually applied when the values of a property θ are known for three meshes with the same grid refinement factor ratio between them [27]. In this case, the Richardson extrapolation is given by:

$$\theta_{ext} = \frac{R^{ap}\theta_2 - \theta_3}{R^{ap} - 1}, \tag{A1}$$

where,

$$ap = \frac{\ln|(\theta_3 - \theta_2)/(\theta_2 - \theta_1)|}{\ln R}, \tag{A2}$$

and then, the error is estimated by:

$$e_i(\%) = \frac{\theta_i - \theta_{ext}}{\theta_{ext}} \times 100, \tag{A3}$$

where θ_{ext} is the variable of interest that we want to extrapolate, the subscripts 3, 2, 1 represent the value of the variable obtained for the three levels of mesh refinement, and ap is the apparent order, given by Equation (A2). R is the grid refinement factor that should be greater than 1.3 [31]. In all case studies presented in this work, we assumed the value of two; consequently, the number of cells in each direction was doubled for consecutive mesh refinement levels [31]. Following a common procedure, in this work, the results were considered to be converged when the relative error, given by Equation (A3), was below 5% [27].

The above relation of Richardson's extrapolation relied on the assumption that the convergence is asymptotic [31]. However, in several cases, this did not happen, and oscillatory convergence was observed for the variables [27]. In those cases, instead of using Equation (A1), a different variation of the method could be used. Assuming that ap was the order of accuracy of the discretization schemes used in the calculation procedure, instead of using three meshes, the extrapolated value could be obtained by Equation (A1) using only the two most refined meshes [32].

References

1. Kennedy, P.; Zheng, R. *Flow Analysis of Injection Molds*, 2nd ed.; Hanser: Munich, Germany, 2001.
2. OpenFOAM. The Open Source CFD Toolbox. Available online: https://www.openfoam.com/ (accessed on 27 May 2020).

3. Habla, F.; Fernandes, C.; Maier, M.; Densky, L.; Ferrás, L.L.; Rajkumar, A.; Carneiro, O.S.; Hinrichsen, O.; Nóbrega, J.M. Development and validation of a model for the temperature distribution in the extrusion calibration stage. *Appl. Therm. Eng.* **2016**, *100*, 538–552. [CrossRef]
4. Rajkumar, A.; Ferrás, L.L.; Fernandes, C.; Carneiro, O.S.; Nóbrega, J.M. Guidelines for balancing the flow in extrusion dies: The influence of the material rheology. *J. Polym. Eng.* **2017**, *38*, 197–211. [CrossRef]
5. Araújo, B.J.; Teixeira, J.C.F.; Cunha, A.M.; Groth, C.P.T. Parallel three-dimensional simulation of injection molding process. *Int. J. Numer. Methods Fluids* **2009**, *59*, 801–815. [CrossRef]
6. Nagy, J.; Steinbichler, G. Fluid Dynamic and Thermal Modeling of the Injection Molding Process in OpenFOAM. In *OpenFOAM Selected Papers of the 11th Workshop*; Nóbrega, J.M., Jasak, H., Eds.; Springer: Cham, Switzerland, 2019.
7. Magalhães, A. OpenFOAM Simulation of the Injection Moulding Filling Stage. Master's Thesis, Integrated Master in Polymer Engineering, University of Minho, Braga, Portugal, September 2016. Available online: http://repositorium.sdum.uminho.pt/handle/1822/65291 (accessed on 27 May 2020).
8. Mole, N.; Krebelj, K.; Stŏk, B. Injection molding simulation with solid semi-crystalline polymer mechanical behavior for ejection analysis. *Int. J. Adv. Manuf. Technol.* **2017**, *93*, 4111–4124. [CrossRef]
9. Krebelj, K.; Turk, J. An Open–Source Injection Molding Simulation. A Solver for OpenFOAM. Available online: https://github.com/krebeljk/openInjMoldSim (accessed on 10 September 2019).
10. Fontaínhas, A. Injection Moulding Simulation Using OpenFOAM. Master's Thesis, Integrated Master in Polymer Engineering, University of Minho, Braga, Portugal, 2019. Available online: https://www.researchgate.net/publication/341448307_Injection_moulding_simulation_using_OpenFOAM (accessed on 27 May 2020).
11. Moldex 3D. Plastic Injection Molding Simulation Software. Available online: https://www.moldex3d.com/ (accessed on 27 May 2020).
12. Pichelin, E.; Coupez, T. Finite element solution of the 3D mold filling problem for viscous incompressible fluid. *Comput. Methods Appl. Mech. Eng.* **1998**, *163*, 359–371. [CrossRef]
13. Berberović, E.; van Hinsberg, N.P.; Jakirlić, S.; Roisman, I.V.; Tropea, C. Drop impact onto a liquid layer of finite thickness: Dynamics of the cavity evolution. *Phys. Rev. E* **2009**, *79*, 036306. [CrossRef] [PubMed]
14. Hirt, C.W.; Nichols, B.D. Volume of fluid (VOF) method for the dynamics of free boundaries. *J. Comput. Phys.* **1981**, *39*, 201–225. [CrossRef]
15. Haider, J. Numerical Modelling of Evaporation and Condensation Phenomena. Master's Thesis, Universität of Stuttgart, Stuttgart, Germany, 2013.
16. Rusche, H. Computational Fluid Dynamics of Dispersed Two-Phase Flows at High Phase Fractions. Ph.D. Thesis, Department of Mechanical Engineering from the Imperial College of Science, Technology and Medicine, University of London, London, UK, 2002.
17. Chiang, H.H.; Hieber, C.A.; Wang, K.K. A unified simulation of the filling and postfilling stages in injection molding. Part I: Formulation. *Polym. Eng. Sci.* **1991**, *31*, 116–124. [CrossRef]
18. Williams, M.L.; Landel, R.F.; Ferry, J.D. The Temperature Dependence of Relaxation Mechanisms in Amorphous Polymers and Other Glass-forming Liquids. *J. Am. Chem. Soc.* **1955**, *77*, 3701–3707. [CrossRef]
19. Zhou, H. *Computer Modelling for Injection Molding*; John Wiley & Sons, Inc.: Hoboken, NJ, USA, 2013.
20. Rees, H. *Understanding Injection Mold Design*; Hanser Publications: Cincinnati, OH, USA, 2001.
21. Tait, P.G. *Physics and Chemistry of the Voyage of H.M.S. Challenger*; HMSO: London, UK, 1888; Volume 2, pp. 941–951.
22. Ferziger, J.H.; Peric, M. *Computational Methods for Fluid Dynamics*; Springer: Cham, Switzerland, 2001.
23. Jasak, H. Error Analysis and Estimation for the finite Volume Method with Application to Fluid Flows. Ph.D. Thesis, Imperial College, University of London, London, UK, 1996.
24. Liang, J.; Luo, W.; Huang, Z.; Zhou, H.; Zhang, Y.; Zhang, Y.; Fu, Y. A robust finite volume method for three-dimensional filling simulation of plastic injection molding. *Eng. Comput.* **2017**, *34*, 814–831. [CrossRef]
25. Kim, K.H.; Isayev, A.I.; Kwon, K.; van Sweden, C. Modelling and experimental study of birefringence in injection molding of semicrystalline polymers. *Elsevier Polym.* **2005**, *46*, 4183–4203. [CrossRef]
26. Rosato, D.V.; Rosato, M.G. *Injection Molding Handbook*; Springer: Cham, Switzerland, 2000.
27. Celik, I.; Ghia, U.; Freitas, C.J.; Coleman, H.; Raad, P.E. Procedure for Estimation and Reporting of Uncertainty Due to Discretization in CFD Applications. *J. Fluids Eng.* **2008**, *130*, 1–4. [CrossRef]
28. Oberkampf, W.L.; Trucano, T.G.; Hirsch, C. Verification, validation, and predictive capability in computational engineering and physics. *Appl. Mech. Rev.* **2004**, *57*, 345–384. [CrossRef]

29. Juretic, F. A Library for Automatic Mesh Generation. Available online: https://cfmesh.com/cfmesh/ (accessed on 27 May 2020).
30. Richardson, L.F. On the Approximate Arithmetical Solution by Finite Differences of Physical Problems Involving Differential Equations, With an Application to the Stresses in a Masonary Dam. *Philos. Trans. R. Soc. Lond.* **1910**, *210*, 307–357. [CrossRef]
31. Celik, I.; Karatekin, O. Numerical experiments on application of Richardson extrapolation with non-uniform grids. *J. Fluids Eng.* **1997**, *119*, 584–590. [CrossRef]
32. Fernandes, C.; Faroughi, S.A.; Carneiro, O.S.; Miguel Nóbrega, J.; McKinley, G.H. Fully-resolved simulations of particle-laden viscoelastic fluids using an immersed boundary method. *J. Non Newton. Fluid Mech.* **2019**, *266*, 80–94. [CrossRef]

© 2020 by the authors. Licensee MDPI, Basel, Switzerland. This article is an open access article distributed under the terms and conditions of the Creative Commons Attribution (CC BY) license (http://creativecommons.org/licenses/by/4.0/).

Article

Antibiotic Activity Screened by the Rheology of *S. aureus* Cultures

Raquel Portela [1], Filipe Valcovo [1], Pedro L. Almeida [2,3], Rita G. Sobral [1] and Catarina R. Leal [2,3,*]

[1] UCIBIO@REQUIMTE, Faculdade de Ciências e Tecnologia, Universidade Nova de Lisboa, 2829-516 Caparica, Portugal; rp.portela@campus.fct.unl.pt (R.P.); f.valcovo@campus.fct.unl.pt (F.V.); rgs@fct.unl.pt (R.G.S.)
[2] Área Departamental de Física, ISEL—Instituto Superior de Engenharia de Lisboa, Instituto Politécnico de Lisboa, 1959-007 Lisboa, Portugal; palmeida@adf.isel.pt
[3] CENIMAT/I3N, Departamento de Ciência dos Materiais, Faculdade Ciências e Tecnologia, Universidade Nova de Lisboa, 2829-516 Caparica, Portugal
* Correspondence: cleal@adf.isel.pt; Tel.: +351-21-831-7135

Received: 17 March 2020; Accepted: 14 May 2020; Published: 18 May 2020

Abstract: Multidrug resistant bacteria are one of the most serious public health threats nowadays. How bacteria, as a population, react to the presence of antibiotics is of major importance to the outcome of the chosen treatment. In this study we addressed the impact of oxacillin, a β-lactam, the most clinically relevant class of antibiotics, in the viscosity profile of the methicillin resistant *Staphylococcus aureus* (MRSA) strain COL. In the first approach, the antibiotic was added, at concentrations under the minimum inhibitory concentration (sub-MIC), to the culture of *S. aureus* and steady-state shear flow curves were obtained for discrete time points during the bacterial growth, with and without the presence of the antibiotic, showing distinct viscosity progress over time. The different behaviors obtained led us to test the impact of the sub-inhibitory concentration and a concentration that inhibited growth. In the second approach, the viscosity growth curves were measured at a constant shear rate of $10\ \text{s}^{-1}$, over time. The obtained rheological behaviors revealed distinctive characteristics associated to the presence of each concentration of the tested antibiotic. These results bring new insights to the bacteria response to a well-known bacteriolytic antibiotic.

Keywords: rheology; MRSA; *S. aureus*; antibiotics; oxacillin; bactericidal

1. Introduction

Bacterial multidrug resistance is the root of the antibiotic crisis that humankind is beginning to face and that will become disastrous in the very near future. In view of the increasing shortage of available therapies that are efficient against bacterial strains that have adapted and developed numerous resistance mechanisms, new strategies are needed. Treatment failure is often the consequence of sub-populations of cells that become resistant or tolerant to the antibiotic pressure. Many parameters influence the outcome, such as presence of other microorganisms, the host immune system or nutrient limitations. Another important factor, often neglected, is the shear stress conditions to which bacteria are submitted, inside and outside the host. To develop new antibacterial strategies, knowledge must be obtained on how the bacteria respond to the presence of antibiotics and under shear stress, as a complex population both as biofilms and in the planktonic state.

It is already well established that the viscoelasticity of biofilms, which critically depends on their structure and composition, plays a major role in their protective effect against mechanical and chemical challenges [1]. In this regard, one can find several works in the literature concerning the characterization of the mechanical behavior of bacterial biofilms, although most are implemented over solid biofilms as they occur in real situations [2–4], to study the adhesion properties to surfaces.

A recent study compared the impact of having complex medium tryptic soy broth supplemented by glucose and NaCl (TGN) or artificial sputum medium (ASM), that mimics cystic fibrosis sputum (CFS), as the nutrient support for biofilm formation, in the response of *S. aureus* to several antibiotics. The authors claimed that the higher elastic properties of ASM, very similar to the ones of CFS, contribute actively towards the observed outcome, that all antibiotics were drastically less efficient in ASM than in TGN towards clearing of *S. aureus* [5].

In our recent works, the mechanical behavior of *S. aureus* planktonic cultures was accessed by biological methods in conjunction with rheological and rheo-imaging techniques [6–10]. When subjected to a shear flow, the bacterial cultures of *S. aureus* disclosed an intricate rheological behavior without a counterpart in biological procedures characterization. More specifically, in steady shear flow, the viscosity value was enhanced during the exponential phase and reverted throughout the late phase to a value similar to the original one, complemented with the equilibrium of the cell population, exhibiting shear rate dependence [7]. The observed decrease in the viscosity is mainly caused by cell deposition [10]. In oscillatory flow, both elastic and viscous moduli presented power-law dependencies whose exponents are functions of the bacteria growth phase and can be concomitant with a soft glassy material response [11]. These behaviors were enclosed in a microscopic model that considers the development of a dynamic web-like structure, where particular aggregation phenomena might arise, according to the growth phase and bacterial density. By merging optical density assays and dry weight determination procedures, recent studies brought new indications confirming that the bacterial aggregation patterns that arise throughout growth, under shear, cannot be attributed uniquely to a cell population density dependence [9].

In this study we addressed the impact of the β-lactam antibiotic oxacillin (Oxa) on the early methicillin resistant *S. aureus* (MRSA) strain COL [12]. MRSA strains emerged in the 1960s, after the introduction of β-lactamase-resistant β-lactam antibiotics, the class of antibiotics most used in the clinic, that inhibit bacterial cell wall synthesis by binding irreversibly to the transpeptidase domain of penicillin-binding proteins (PBPs) [13]. PBP proteins catalyze the last steps of peptidoglycan biosynthesis, in the external side of the membrane, being responsible for the high cross-linking level of this cell wall macromolecule. The MRSA strains, that are intrinsically resistant to the β-lactam mode of action, emerged through the acquisition of a mobile genetic element, the staphylococcal cassette chromosome *mec* (SCC*mec* cassette), the genetic element carrying the β-lactam resistance gene *mecA*, from other bacterial species [14]. The acquisition of an SCC*mec* cassette occurred in numerous independent events, resulting in distinct MRSA clonal lineages, but all SCC*mec* cassettes contain the *mecA* gene, the central element of resistance. The *mecA* gene encodes an extra PBP for PBP2A, with low affinity for virtually all β-lactam antibiotics, resulting in a lower efficiency of acylation that takes over cell wall synthesis in the presence of β-lactam antibiotics [15]. As for other β-lactam antibiotics, the mode of action of oxacillin is described as being bactericidal, since by inactivating PBPs, the cell wall weakens and the cell eventually lyses [16]. The MRSA strain COL presents a high minimum inhibitory concentration (MIC) value of 400 µg/mL [17]. The MIC corresponds to the lowest concentration of an antibacterial drug that inhibits the growth of bacteria.

The antibiotic oxacillin was added to the cultures of the *S. aureus* strain, in different concentrations and different rheological experimental approaches were applied. Namely, steady-state flow curves and viscosity growth curve measurements. Moreover, the viscosity growth curves were obtained at a constant shear rate value over time, compatible with human physiological values [18]. The obtained rheological behaviors revealed distinctive characteristics associated with the presence of the tested antibiotic. These differences are justified by the antibiotic bacteriolytic mode of action.

2. Materials and Methods

2.1. Bacterial Strain and Growth Conditions

The methicillin resistant *Staphylococcus aureus* (MRSA) strain COL, a gram-positive human pathogen [19] was used. The strain was grown overnight in tryptic soy agar medium (TSA) plates at 37 °C. A single colony of the strain was inoculated in 5 mL of fresh tryptic soy broth (TSB) and grown overnight at 37 °C at 180 rpm to promote the aeration of the culture.

Cultures were grown at 37 °C with aeration in an orbital shaker (180 rpm) and the optical density at 620 nm ($OD_{620\,nm}$) was monitored over time using a spectrophotometer (Ultrospec 2100 pro).

To determine the most appropriate concentrations of antibiotic to study, growth curves were performed in the presence of oxacillin at lower and higher concentrations than the MIC value (200, 400, 800 and 1600 µg/mL) added to the culture at 180 min of growth.

The influence of the antibiotic in the growth process of the culture was assessed considering two approaches regarding rheological characterization: A, Addition of antibiotic at a specific time of growth, namely at t = 300 min, to the culture while still in the incubator, followed by the rheological measurement; B, addition of antibiotic at specific time points of growth during the rheological measurement. In both approaches *S. aureus* cultures were set by inoculating fresh TSB medium with the starting cultures to obtain an initial $OD_{620\,nm} = 0.005$.

Approach A:

At ~300 min of growth, 200 µg/mL of oxacillin was added to the culture in the flask. A sample was taken to continue growth in the rheometer. After the addition of the antibiotic the cultures remained in the incubator with the $OD_{620\,nm}$ monitored. At time points, 1, 2 and 3 h of incubation, samples of the cultures were taken in triplicate to determine the cell viable counts (CFU/mL) by plating serial dilutions of the bacterial cultures on TSA medium without antibiotic. For the same time points, a calibrated volume of sample was collected and observed by optical microscopy using an Olympus BX50 microscope with an Olympus SC30 camera and AnalySIS getIT 5.1 image software. For each aliquot, an average of 10 photos were taken randomly. This optical analysis was also applied to the concentration of oxacillin of 800 µg/mL.

Approach B:

The cultures were loaded on the rheometer at an $OD_{620\,nm}$ of 0.005 and grown at 37 °C under shear. At ~340 min of growth, oxacillin was added to a final concentration of 200 and 800 µg/mL. This procedure was carefully performed to minimize the interference in the running measurement, such as to prevent contact with the rotating upper geometry and to assure the dispersion of the antibiotic inside the sample.

2.2. Rheology

Rheological measurements were performed in a controlled stress rotational rheometer Bohlin Gemini HRnano, and different geometries were used to perform two types of assays:

(i) flow curve: a steel cone and plate geometry, with diameter 40 mm, angle 2° and gap 55 µm for the steady-state shear flow tests, which were performed at 20 °C. Each step was acquired assuring that a minimum of 1500 units of deformation was imposed to the sample. Using a CP geometry with a very small gap, sample ejection, due to centrifuge effect, did not occur within the range of shear rate values considered;

(ii) viscosity growth curve: a steel plate-plate geometry, with diameter 40 mm and gap of 2000 µm (to ensure a good signal) for the steady shear viscosity measurement, imposing a constant shear rate of 10 s^{-1}, during time and at 37 °C to allow optimal growth conditions.

In approach B tests, the addition of the antibiotic was performed using a micropipette with a tip diameter of 0.5 mm which is 4 times smaller than the gap used in the measurement. For control purposes we have previously used an acrylic PP geometry, to follow the addition of a dye to the sample, in the same volume as the antibiotic, which gave us an insight on how the dispersion occurs, and no peripherical droplet accumulation was verified.

Assays were performed in triplicate over fresh cultures and a solvent trap was used to minimize evaporation.

3. Results and Discussion

3.1. Antibiotic Effect on S. aureus Growth

Previous studies showed that the oxacillin MIC value of strain COL in solid medium was 400 µg/mL [17]. To study the rheological effect of the challenge of oxacillin on an MRSA strain, we first determined the most appropriate concentrations of antibiotic to use, lower and higher than the MIC value (200, 400, 800 and 1600 µg/mL). The optical density of a COL culture, in function of the added antibiotic concentration, was measured over growth and, as expected, the growth rate of the population challenged with increasing concentrations of oxacillin, decreased accordingly, see Figure 1a.

We chose the concentration of 200 µg/mL, corresponding to half the MIC value to perform a first set of experiments, steady state shear flow tests. This concentration of oxacillin resulted in a slower growth rate, during the exponential phase, but the cell division process was clearly still active, as observed by optical density monitoring, see Figure 1a. For further rheological studies, we also chose the concentration of 800 µg/mL, corresponding to twice the MIC value. This concentration of oxacillin resulted in an arrest in growth, approximately 100 min after antibiotic addition (see Figure 1b), followed by a decrease in the optical density of the culture, suggesting arrest of cell division associated with cell lysis.

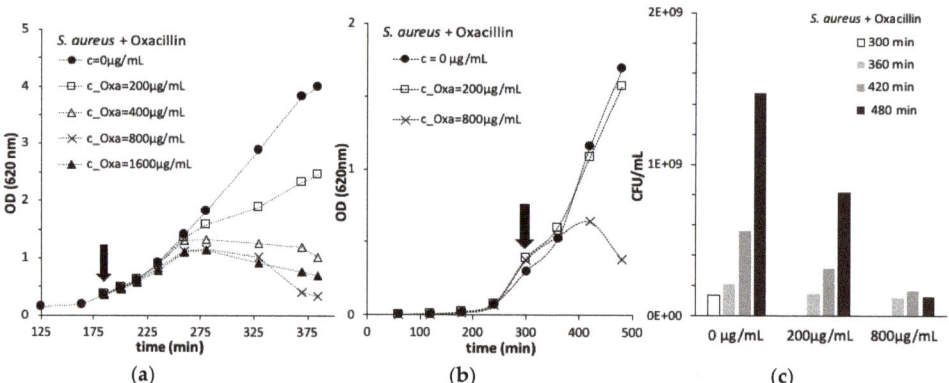

Figure 1. (a) Growth curves of *S. aureus* strain COL growth monitored by measuring the optical density ($OD_{620\,nm}$) over time. *S. aureus* was grown in the absence and in the presence of different concentrations of oxacillin (Oxa), at 200, 400, 800 and 1600 µg/mL added at t = 180 min (representative curves). The arrow represents the time of addition of the antibiotic to the culture. (b) Growth curves of *S. aureus* strain COL grown in the absence and in the presence of different concentrations of antibiotics: oxacillin (Oxa) at 200 and 800 µg/mL added at 300 min (representative curves). The arrow represents the time of addition of the antibiotic to the culture. (c) Colony forming units (CFU/mL), at specific growth times: 300, 360, 420 and 480 min after the addition of oxacillin. All the measurements were performed at 37 °C.

3.2. Steady State Shear Flow Assays

Steady state shear flow tests were performed at 20 °C (to minimize bacteria growth during the assay) and allowance made for the characterization of the viscosity of the cultures, in the absence and in presence of a sub-inhibitory concentration of oxacillin (200 µg/mL) in function of the shear rate.

In general, the flow curves revealed a shear-thinning behavior followed by a shear-thickening behavior for the *S. aureus* cultures, with and without the presence of antibiotics. In Figure 2a the flow curves for the *S. aureus* cultures with the addition of oxacillin are plotted for different growth times (for *S.aureus* flow curves please see Figure 2 in [6]). For samples collected at the beginning of the $OD_{620\ nm}$ growth curve (see Figure 1b), the viscosity appears to be smaller than the culture medium viscosity, Figure 2a. Only for intermediate and longer growth times, 395 min and 485 min, the viscosity shows higher values than the culture medium viscosity, for shear rate values lower than 300 s^{-1}, in the shear-thinning region. For higher shear rate values, above 600 s^{-1}, all the flow curves tend into the same curve with almost the same slope of the culture medium.

The representation of the viscosity values obtained from the flow curves for the shear-rate value of 100 s^{-1}, in function of time, Figure 2b, showed that in the presence of oxacillin, the viscosity of the culture almost doubled when compared with the viscosity of the culture in the absence of antibiotic, that specifically occurred during the exponential phase, in the time range ~450–550 min. This obvious increase in viscosity may be associated with the release of cytoplasmic components to the extracellular environment due to lysis of cells, due to the bacteriolytic action of oxacillin.

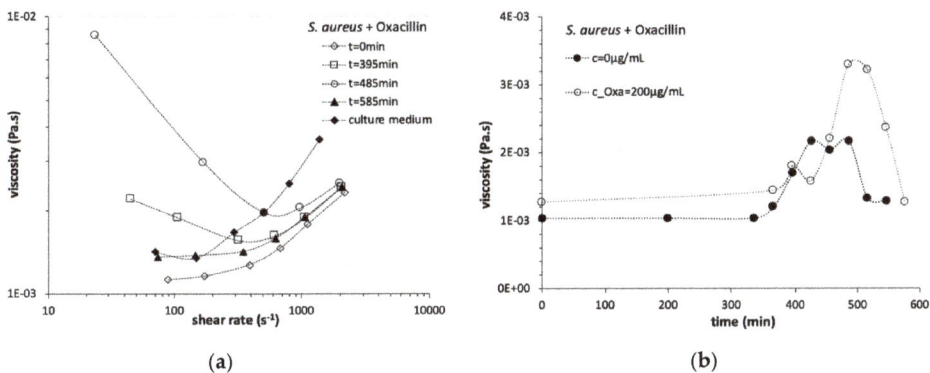

Figure 2. Steady-state shear flow characterization for bacterial cultures of *S. aureus*, with and without the addition of antibiotic during bacterial growth in the incubator at t = 350 min. Approach A: (**a**) flow curves of culture aliquots, removed from the incubator at specific time points during culture growth at: t = 0 min (open diamond symbol), t = 395 min (open square symbol), t = 485 min (open circle symbol), t = 585 min (full triangle symbol), culture medium (full black diamond symbol) and lines are guide to the eye. Similar behaviors were obtained for cultures without addition of oxacillin (representative curves). (**b**) Representation of shear viscosity values in function of time, obtained from the flow curves for the shear rate value of 100 s^{-1}: without antibiotic (full black symbol) and with oxacillin at c_Oxa = 200 µg/mL (full gray symbol). All measurements were performed at 20 °C to minimize cell growth during the assays.

3.3. Viscosity Growth Curves

The viscosity growth curves were determined for the two concentrations of oxacillin, added inside or outside of the rheometer (in the culture flask) after approximately 300 min of growth. For three consecutive 1 h separate points, cell viable counts were determined, and the cultures were observed and compared by optical microscopy. The images obtained, Figure 3, show the increase of cell density over time in the *S. aureus* culture and the effect of the presence of the antibiotic in the growth process.

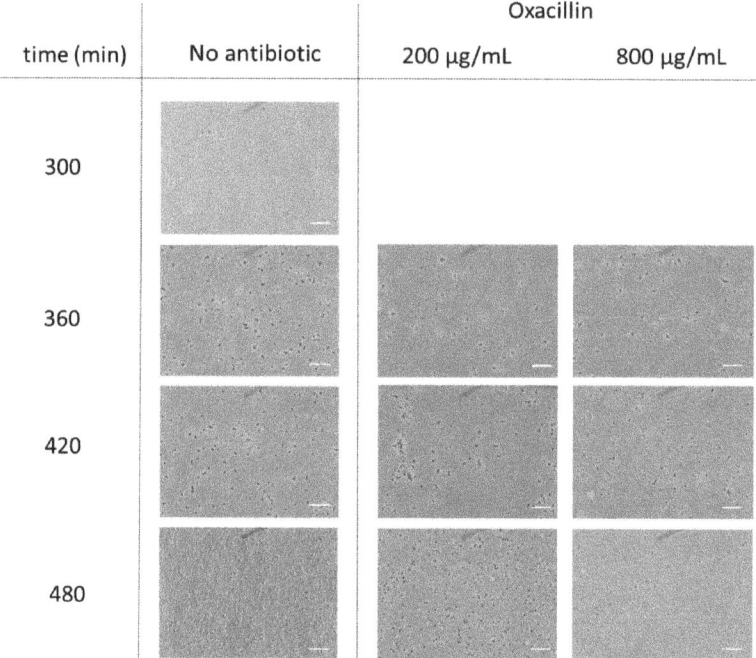

Figure 3. Optical microscopy images of *S. aureus* strain COL grown in the absence (left panel) and in the presence of different concentrations of oxacillin (Oxa) at 200 and 800 μg/mL (right panel), added at 300 min of growth. For each aliquot a calibrated volume of sample of 5 μL was collected and observed. An average of 10 photos were taken randomly at specific growth times: 0, 300, 360, 420 and 480 min. *S. aureus* are coccoid shaped cells with an average diameter of 0.5–1 μm. Scale bar represents 10 μm width.

For $c\_Oxa = 200$ μg/mL, this condition did not alter the optical density profile, suggesting that the rate of cell division is maintained, see Figure 1b. However, this behavior showed no counterpart in the viability profile obtained from the progress of the CFUs/mL, see Figure 1c. Although maintaining a continually increasing pace, a decrease in the viability counts of approximately 40%, with respect to the culture with no antibiotic addition, was observed at 480 min, suggesting a slowdown in cell division. This discrepancy may be explained by a cell aggregation effect of oxacillin at sub-MIC concentrations, which was in fact observed by optical microscopy at 120 and 180 min after antibiotic addition. The principle of the CFUs measurement is that a single cell, when positioned in the agar plate will originate a single colony. When bacteria are growing in the liquid medium aggregated in groups comprising a high number of cells together, when performing the CFUs, the aggregate will be maintained in the agar plate and originate a single colony. If bacteria are growing in a non-aggregated form and instead all the cells in liquid medium are growing separately, when the medium is plated in agar to perform the CFUs, the same number of cells will originate a higher number of colonies.

For the higher concentration $c\_Oxa = 800$ μg/mL, the expected decrease in optical density was observed at approximately 120 min after antibiotic addition, see Figure 1b, meaning a drop in cell division rate. More clearly, from the viability profile, cell division is shown to be arrested from 420 min onwards, as shown in Figure 1c. A decrease in the viability counts of approximately 90%, with respect to the culture with no antibiotic addition, was observed for this double MIC concentration. This behavior is in accordance with a cell lysis scenario, associated with the bactericidal mode of action of oxacillin, and is corroborated by the fact that virtually no cells can be observed by optical microscopy at 480 min.

To directly explore the influence of the antibiotic presence in the rheological behavior of *S. aureus* cultures, during growth process, two different approaches were followed, by measuring the viscosity growth curve, see Figures 4 and 5. The viscosity growth curve, at sub-inhibitory antibiotic concentration follows the previously described behavior, which was framed by a microscopic model [7]. According to this model, at different growth stages, since the density of bacteria increases, the bacterial cells rearrange themselves in aggregates forming dynamic web-structures, triggering different viscoelastic responses. In these cases, the addition of the antibiotic occurred at 300 min in the incubator, approach A, after which the culture was placed in the rheometer and the viscosity growth curve measurement initiated. The cultures of *S. aureus*, with and without the addition of antibiotic, showed an almost immediate sharp increase in the viscosity, as previously reported [6] for the last case, since the exponential phase of growth initiates by this time, see Figure 4. Comparable maximum values of the viscosity were attained by each culture, corresponding to ~30× the initial value, moreover the decay in the viscosity occurred in an abrupt way and happened in ~300 min, when no antibiotic was present in the culture, while in the culture with oxacillin the viscosity decreased slowly and took almost three times as long to recover the initial viscosity value.

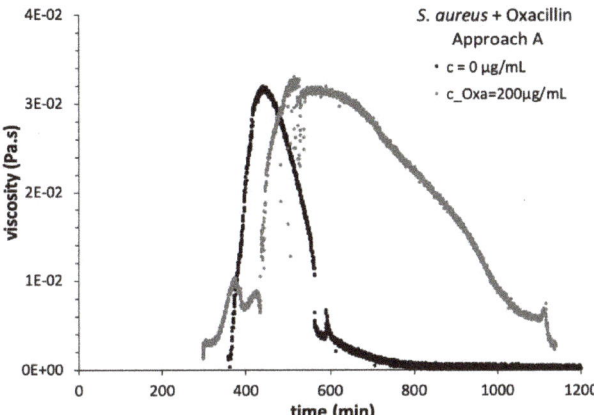

Figure 4. Viscosity growth curves for bacterial cultures of *S. aureus*, with and without the addition of antibiotic in the incubator at t = 300 min, approach A, followed by rheologic measurement: with no addition of antibiotic c = 0 µg/mL (full black symbol) and with oxacillin in concentration c_Oxa = 200 µg/mL (full gray symbol); all measurements were performed at a constant shear rate of $10\ s^{-1}$ (representative curves) and at 37 °C.

In Figure 5, the viscosity growth curve was measured starting with the culture of *S. aureus* and adding the antibiotic in each respective concentration, directly to the culture in the rheometer, during the assay, in the beginning of the exponential phase, approach B. For the lower antibiotic concentration, a similar behavior to the one verified with the previous addition procedure, approach A (see Figure 4), was obtained, with slight differences. The viscosity growth curves obtained with the addition of the highest antibiotic concentration showed the incapacity of the cultures to grow in such conditions, as anticipated.

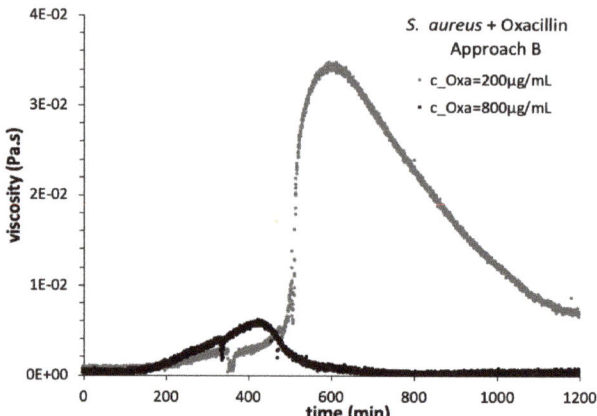

Figure 5. Viscosity growth curves for bacterial cultures of *S. aureus*, with the addition of the antibiotic oxacillin directly to the sample in the rheometer, approach B. The bacteriolytic antibiotic oxacillin was added at t = ~340 min for two concentrations: c_Oxa = 200 μg/mL (full gray symbol) and c_Oxa = 800 μg/mL (full black symbol); all measurements were performed at a constant shear rate of $10\,\text{s}^{-1}$ (representative curves) and at 37 °C.

4. Conclusions

In this study we addressed the impact of oxacillin, a β-lactam antibiotic, in the viscosity profile of the methicillin resistant *S. aureus* (MRSA) strain COL, through in-situ rheology applied during culture growth.

The addition of oxacillin, to the *S. aureus* cultures, at sub-inhibitory concentrations induced a more pronounced increase in the viscosity values during the exponential phase of growth. Several mechanisms may be contributing to this behavior, including the observed cell aggregation, partial cell lysis and the release of cell content. Finally, another hypothesis related to the secretion of compounds to the extracellular environment occurring when the bacterial cells are challenged with the antibiotic, cannot be ruled out at this moment.

For the inhibitory concentration, as expected, the cell division was rapidly arrested and as a result the rheological behaviors were drastically affected.

As future work, we also aim to explore bacterial growth in different environments and challenges with other antibiotic classes.

Author Contributions: Conceptualization, R.P., P.L.A., R.G.S. and C.R.L.; methodology, R.P., F.V., P.L.A., R.G.S. and C.R.L.; writing—original draft preparation, C.R.L.; writing—review and editing, R.P., P.L.A., R.G.S. and C.R.L. All authors have read and agreed to the published version of the manuscript.

Funding: This work was supported by FEDER through COMPETE 2020; FCT, Portuguese Foundation for Science and Technology Projects No. UID/CTM/50025/2019 (CENIMAT) and PTDC/FIS-NAN/0117/2014 (P.L.A.); PTDC/BIA-MIC/31645/2017 (R.G.S.); and the Applied Molecular Biosciences Unit, UCIBIO, which is financed by national funds from FCT (UIDB/ 04378/2020) and co-financed by the ERDF under the PT2020 Partnership Agreement No. POCI-01-0145-FEDER-007728.

Conflicts of Interest: The authors declare no conflict of interest.

References

1. Peterson, B.W.; He, Y.; Ren, Y.; Zerdoum, A.; Libera, M.R.; Sharma, P.K.; van Winkelhoff, A.J.; Neut, D.; Stoodley, P.; van der Mei, H.C.; et al. Viscoelasticity of biofilms and their recalcitrance to mechanical and chemical challenges. *FEMS Microbiol. Rev.* **2015**, *39*, 234. [CrossRef] [PubMed]
2. Klapper, I.; Rupp, C.J.; Cargo, R.; Purevdorj, B.; Stoodley, P. Viscoelastic fluid description of bacterial biofilm material properties. *Biotech. Bioeng.* **2002**, *80*, 289. [CrossRef] [PubMed]

3. Towler, B.W.; Rupp, C.J.; Cunningham, A.B.; Stoodley, P. Viscoelastic properties of a mixed culture biofilm from rheometer creep analysis. *Biofouling* **2003**, *19*, 279. [CrossRef] [PubMed]
4. Rogers, S.S.; van der Walle, C.; Waigh, T.A. Microrheology of Bacterial Biofilms In Vitro: *Staphylococcus aureus* and *Pseudomonas aeruginosa*. *Langmuir* **2008**, *24*, 13549. [CrossRef] [PubMed]
5. Iglesias, Y.D.; Wilms, T.; Vanbever, R.; Bambekea, F.V. Activity of antibiotics against *Staphylococcus aureus* in an in vitro model of biofilms in the context of cystic fibrosis: Influence of the culture medium. *Antimicrob. Agents Chemother.* **2019**, *63*, e00602. [CrossRef] [PubMed]
6. Portela, R.; Almeida, P.L.; Patrício, P.; Cidade, T.; Sobral, R.G.; Leal, C.R. Real-time rheology of actively growing bacteria. *Phys. Rev. E* **2013**, *87*, 030701(R). [CrossRef]
7. Patrício, P.; Almeida, P.; Portela, R.; Sobral, R.; Grilo, I.; Cidade, T.; Leal, C.R. Living bacteria rheology: Population growth, aggregation patterns, and collective behavior under different shear flows. *Phys. Rev. E* **2014**, *90*, 022720. [CrossRef] [PubMed]
8. Portela, R.; Patrício, P.; Almeida, P.L.; Sobral, R.G.; Franco, J.M.; Leal, C.R. Rotational tumbling of Escherichia coli aggregates under shear. *Phys. Rev. E* **2016**, *94*, 062402. [CrossRef] [PubMed]
9. Portela, R.; Almeida, P.L.; Sobral, R.G.; Leal, C.R. Motility and cell shape roles in the rheology of growing bacteria cultures. *Eur. Phys. J. E* **2019**, *42*, 26. [CrossRef]
10. Franco, J.M.; Patrício, P.; Almeida, P.; Portela, R.; Sobral, R.G.; Leal, C.R. Cell necklaces behave as a soft glassy material. In Proceedings of the Ibereo 2015 Conference—Challenges in Rheology and Product Development, Coimbra, Portugal, 7–9 September 2015; Rasteiro, M.G., Ed.; pp. 118–121, ISBN 978-989-26-1056-6.
11. Sollich, P.; Lequeux, F.; Hebraud, P.; Cates, M.E. Rheology of Soft Glassy Materials. *Phys. Rev. Lett.* **1997**, *78*, 2020. [CrossRef]
12. Jevons, M.P. "Celbenin" - resistant Staphylococci. *Br. Med. J.* **1961**, *1*, 124–125. [CrossRef]
13. Ghuysen, J.M. Molecular structures of penicillin-binding proteins and β-lactamases. *Trends Microbiol.* **1994**, *2*, 372–380. [CrossRef]
14. Miragaia, M. Factors Contributing to the Evolution of *mecA*-Mediated β-lactam Resistance in Staphylococci: Update and New Insights from Whole Genome Sequencing (WGS). *Front. Microbiol.* **2018**, *9*, 2723. [CrossRef] [PubMed]
15. Fuda, C.; Suvorov, M.; Vakulenko, S.B.; Mobashery, S. The Basis for Resistance to b–Lactam Antibiotics by Penicillinbinding Protein 2a of Methicillin-resistant *Staphylococcus aureus*. *J. Biol. Chem.* **2004**, *279*, 40802–40806. [CrossRef] [PubMed]
16. Raynor, R.H.; Scott, D.F.; Best, G.K. Oxacillin-induced lysis of Staphylococcus aureus. *Antimicrob Agents Chemother.* **1979**, *16*, 134–140. [CrossRef] [PubMed]
17. Sobral, R.G.; Ludovice, A.M.; Gardete, S.; Tabei, K.; De Lencastre, H.; Tomasz, A. Normally functioning murF is essential for the optimal expression of methicillin resistance in Staphylococcus aureus. *Microb. Drug Resist.* **2003**, *9*, 231–241. [CrossRef] [PubMed]
18. Paszkowiak, J.J.; Dardik, A. Arterial wall shear stress: Observations from the bench to the bedside. *Vasc. Endovasc. Surg.* **2003**, *37*, 47–56. [CrossRef] [PubMed]
19. Gill, S.R.; Fouts, D.E.; Archer, G.L.; Mongodin, E.F.; DeBoy, R.T.; Ravel, J.; Paulsen, I.T.; Kolonay, J.F.; Brinkac, L.; Beanan, M.; et al. Insights on evolution of virulence and resistance from the complete genome analysis of an early methicillin-resistant Staphylococcus aureus strain and a biofilm-producing methicillin-resistant Staphylococcus epidermidis strain. *J. Bacteriol.* **2005**, *187*, 2426–2438. [CrossRef] [PubMed]

© 2020 by the authors. Licensee MDPI, Basel, Switzerland. This article is an open access article distributed under the terms and conditions of the Creative Commons Attribution (CC BY) license (http://creativecommons.org/licenses/by/4.0/).

Article

Gravitational Effects in the Collision of Elasto-Viscoplastic Drops on a Vertical Plane

Cassio M. Oishi [1,*], Fernando P. Martins [1] and Roney L. Thompson [2]

[1] Departamento de Matemática e Computação, Faculdade de Ciências e Tecnologia, Universidade Estadual Paulista "Júlio de Mesquita Filho", Presidente Prudente, SP 19060-900, Brazil; fp.martins@unesp.br

[2] Department of Mechanical Engineering, COPPE, Universidade Federal do Rio de Janeiro, Centro de Técnologia, Ilha do Fundão, Rio de Janeiro, RJ 24210-240, Brazil; rthompson@mecanica.coppe.ufrj.br

* Correspondence: cassio.oishi@unesp.br

Received: 17 March 2020; Accepted: 22 April 2020; Published: 27 April 2020

Abstract: The collision of drops in a solid substrate is an interesting problem with several practical applications. When the drop is made of a complex fluid the problem presents numerical challenges due to the interaction of the mechanical properties and the free surface approach. In the present work, we solve the numerical problem of elasto-viscoplastic drops colliding in vertical plane. The free surface evolution is handled by a Marker-And-Cell method combined with a Front-Tracking interface representation. Special emphasis is given to the gravitational effects by means of exploring the Froude number. We were able to find a rich variety of outputs that can be classified as sticking, sliding, bouncing, detaching, and slithering.

Keywords: drop impact; elasto-viscoplastic material; free surface; gravitational effects

1. Introduction

The study of liquid drops on a solid substrate is an important problem due to the number of industrial applications such as spray-cooling of hot surfaces, inkjet printing, agriculture pesticides, spray coating, precision molten drop deposition, fire suppression, internal combustion engines, and others. Depending on the application considered, the focus of the investigation can be very different. While in some studies, the collision of the drop is a relevant part of the analysis, the cases where the drop is deposited on the substrate have also received considerable attention in the literature. From the drop perspective, the analyses can deal with drops made of simple and/or complex fluids, spherical/non-spherical drops, the action of surfactants in the surface tension. With respect to the substrate, variations come also from multiple sources. The surface solid can have different levels of roughness, be more hydrophobic or hydrophilic, be placed with different inclinations with respect to the drop velocity and gravity.

Experimental works of the Newtonian drop impact are mainly focused on the surface tension (e.g., [1–4]) and dynamic contact angle effects (e.g., [5–8]). The motion of the triple point, where the liquid, the solid and the air are in contact is one of the main challenges in droplet dynamics. The oblique drop impact was studied by [9–12] due to the importance of this condition in real applications, mainly establishing conditions where different outputs were obtained. A number of works (e.g., [13,14]) are dedicated to the study of coating treatments and other surface effects that can change the adherence of the solid substrate and hence, the evolution dynamics of the drops that are in contact with this solid.

In the cases where the motion of drops on inclined surfaces is induced by gravity and there is no previous drop collision, the main concerns generally encompass the drop shape and the conditions that induce detachment, rolling and breakage of the drop integrity [15]. The pioneering works conducted by Dussan & Chow [16], Dussan [17] established the firsts fundamental aspects of this problem, examining droplets attached to non-horizontal surfaces and determining conditions for these droplets to slide by means of asymptotic methods. Dimitrakopoulos & Higdon [18] considered the influence of the advancing and the receding contact angles of three-dimensional fluid droplets from inclined solid surfaces. They employed a procedure to optimize the shape of drop seeking for the contact line which yields the maximum displacing force for which the droplet can remain attached to the surface. They were able to recover the results obtained by Dussan & Chow [16], Dussan [17] as asymptotic limits. The experimental study presented by Podgorski et al. [19] analyzed the effect of the inclination angle from a horizontal to a vertical configuration and highlighted the influence of the capillary (ratio of viscous to surface tension effects) and Bond (ratio of gravitational to surface tension effects) numbers on the analysis. Since there is no impact velocity, the initial kinetic energy of the system is null and, therefore, surface tension can act to preserve integrity and to restrict deformation. On the other hand, gravity induces non-symmetric shapes where a "belly" can be formed in the lower part of the drop. Some numerical investigations tried to address this competition between gravity and surface tension, such as the studies conducted by Milinazzo & Shinbrot [20] considering a simplified configuration of the problem and Schwartz et al. [21] who were able to find regimes where the drop splits into smaller ones, as well as a long stretching motion on the vertical wall.

The use of pesticide for pest control in crops is an application where these effects can be important, due to the small size of the droplets that are sprayed on the plants (e.g., [22]). Different contact angles for advancing and receding contact line were considered by Glass et al. [23]. Local variations in wettability in the leaves [24] can be challenging for modelling. In inclined leaves the competition between gravity and surface tension translated by the Bond number plays an important role, as analyzed by Veremiev et al. [25], who also considered the air effect on the determination of the final impact velocity.

Applications where there is a temperature difference between the drop and the wall were also the subject of attention of several works. Fire suppression [26] is a challenging problem where temperature effects need to be considered when the drop hits a wall. Fine-tube heat exchangers applied in refrigeration systems can induce surface low temperatures that induce the formation of water drops at tube walls, as analyzed by Zhuang et al. [27]. While Duy & Vu [28] analyzed liquid drop solidifying on a vertical cold wall using numerical simulations, Demidovich et al. [29] investigated the impact of single and multicomponent liquid drops on a heated wall.

There are important experimental works in the literature that have explored the motion of drops made of complex fluids. Few experimental works examined the motion of a non-Newtonian drop on an inclined plane. Xu et al. [30] investigated this problem with viscoelastic Boger fluids (PEG–PEO solutions) and concluded that elasticity did not significantly influence the results when compared to the Newtonian outputs. Yield stress materials were studied by Jalaal et al. [31] who found significant slip of spreading drops made of Carbopol and of Xanthan gum solutions. The drop impact on a thin liquid film was studied in [32,33] where the limits between sticking and splashing were established using a reduced set of dimensionless numbers where the Reynolds number and the aspect ratio between the drop diameter and the thickness of the liquid played a major role.

Some experimental studies of the drop impact of viscoelastic materials were carried out by [34–38]. As a general conclusion, there are elastic effects of the fluid as whole that replace the contribution of the surface tension in purely Newtonian drops. In the case of yield stress materials, the studies performed by Luu & Forterre [39], Nigen [40], German & Bertola [41], Saidi et al. [42] deserve to be highlighted. The main concern of these works was to investigate the influence of the yield stress and surface tension. The exception is the work of Luu & Forterre [39], where an elasto-viscoplastic analysis was made, since they found some influence of elasticity in the presence of the yield stress.

To address the drop dynamics in these problems from a numerical perspective, a technique to deal with the free surface is necessary. There are some challenges to tackle this problem, even in the Newtonian case. In the non-Newtonian scenario different methods and different fluids were employed in the recent years [43–48], as described by Oishi et al. [49].

Concerning the mechanical properties of the fluid, a broad study on drop impact was conducted in [49], where elasticity, viscoplasticity, inertia, and thixotropy were investigated for large ranges of the corresponding dimensionless numbers. The main outputs considered in the normal impact were bouncing and non-bouncing. The main conclusions with respect to the normal drop impact on a solid surface are summarized next. In general, high levels of the yield stress inhibit bouncing due to highly plastic deformation induced, which is unable to recover. When elasticity is significant, in most of the times, the retraction does not imply on the detachment of the drop from the substrate. Since in viscoelastic fluids (in contrast to viscoelastic solids) higher elastic effects are accomplished by increasing the flexibility of the fluid, the spreading stage is accompanied by large deformations, leading to flatter configurations that have to face a long dissipative path to return to a more rounded configuration and to abandon the solid substrate. Inertia has also a tendency to inhibit bouncing. The initial kinetic energy is used to spread the drop and is dissipated in the spreading and retraction stages (in the case where elasticity is low and surface tension effects are negligible there is no retraction stage, since the fluid is unable to store energy). The influence of surface tension was analyzed in [50] in a subsequent work where it was found that high yield stress inhibits the role of capillary effects, as already demonstrated by Luu & Forterre [39] in their experiments. Again, some results generally attributed to surface tension in inelastic fluids were obtained by an elastic fluid neglecting surface tension effect.

The objective of present work is to investigate the collision of an elasto-viscoplastic drop on a vertical wall. As discussed above, even when the drop is deposited on the wall, gravitational forces lead to an asymmetry in the spreading stage that is worthy of investigation. The impact introduces an addition concern which is the possibility of providing different levels of energy to the system by means of the initial kinetic energy before impact. In addition, sliding is a possible outcome not present in horizontal drop collision. Hence, the study conducted here aims to understand how gravitational effects interact with the mechanical properties of a complex fluid is this problem. In this regard, the role of the Froude number receives special attention and is the main focus of our analysis.

The rest of the text is organized as follows. In Section 2, the governing equations, the model and the numerical scheme are presented. The description of the problem is shown in Section 3. The results are presented in Section 4, where real and parametric mechanical properties are given as input to the problem. Finally, in Section 5, we summarize and draw conclusions with respect to the problem.

2. Governing Equations and Overview of the Numerical Scheme

The mathematical modelling used in this work to describe the complex fluid flows is based on the elasto-viscoplastic framework. Below, we present the motion and constitutive equations in a dimensionless form, following [51,52]:

$$\nabla \cdot \mathbf{u} = 0, \tag{1}$$

$$\frac{\partial \mathbf{u}}{\partial t} + \nabla \cdot (\mathbf{u}\mathbf{u}) = -\nabla p + \frac{1}{Re}\left(\eta_\infty \nabla^2 \mathbf{u} + \nabla \cdot \boldsymbol{\tau}^M\right) + \frac{1}{Fr^2}\mathbf{g}, \tag{2}$$

$$\frac{\partial \boldsymbol{\tau}^M}{\partial t} + \nabla \cdot \left(\mathbf{u}\boldsymbol{\tau}^M\right) - \left[(\nabla \mathbf{u}) \cdot \boldsymbol{\tau}^M + \boldsymbol{\tau}^M \cdot (\nabla \mathbf{u})^T\right] = \frac{f(\lambda)}{Wi}\left[-\frac{1}{\eta(\lambda)}\boldsymbol{\tau}^M + 2\mathbf{D}\right]. \tag{3}$$

In these equations, the velocity and pressure fields are represented by $\mathbf{u}$ and p, respectively, while $\boldsymbol{\tau}^M$ is the non-Newtonian contribution of the total stress tensor which is given by:

$$\sigma = -p\mathbf{I} + 2\eta_\infty \mathbf{D} + \boldsymbol{\tau}^M \tag{4}$$

with $\mathbf{D}$ being the rate of deformation tensor. The dimensionless parameters of Equations (2) and (3) are the Reynold number Re, the viscosity related to the fully unstructured material η_∞, the Froude number Fr and the Weissenberg number Wi. These dimensionless numbers related to inertia, gravity, and elasticity of the fluid are defined by

$$Re = \frac{\rho U^2}{\tau_c}, \quad Fr = \frac{U}{\sqrt{gD}}, \quad Wi = \frac{\tau_c}{G_0}, \tag{5}$$

where U is the characteristic velocity, chosen to be the impact velocity and D is the characteristic length chosen to be the initial drop diameter. The characteristic stress τ_c is chosen to be the flow curve stress value associated with the characteristic shear rate of the problem which in the present case is $\dot{\gamma}_c = U/D$.

In the above equations, the mechanical properties of viscosity and elastic modulus are dependent on a dimensionless structure parameter λ. In a more general approach, we could introduce a time-dependency aspect to the material, like thixotropy for example, by solving a transport equation for λ. However, we explore the elasto-viscoplastic nature of the fluid in this work and, therefore, the microstructure is always at equilibrium conditions and no thixotropic effects are present. In other words, although the mechanical properties depend on the structure parameter, abrupt changes in stress are accompanied with instantaneous changes in microstructure, although elasticity still induces memory effects on the fluid. The material properties $f(\lambda)$ and $\eta(\lambda)$ are defined by

$$f(\lambda) = \frac{G(\lambda)}{G_0} = \exp\left[m\left(\frac{1}{\lambda} - 1\right)\right], \tag{6}$$

and

$$\eta(\lambda) = \left[\left(\frac{\eta_0}{\eta_\infty}\right)^\lambda - 1\right]\eta_\infty, \tag{7}$$

respectively, where η_0 and G_0 are the values of the viscosity and the elastic modulus in the fully structured state, $f(\lambda)$ is a dimensionless function that represents the ratio between the current and fully structured elastic modulus, and m is a positive constant.

In the present model, the flow curve is an input that is used to compute the viscosity at equilibrium conditions and the corresponding structure level. The viscoplastic nature of the material is captured by the presence of a yield stress in the flow curve which is given by

$$\tau = \zeta(\dot{\gamma})\left(\tau_y + K\dot{\gamma}^n\right) + \eta_\infty \dot{\gamma}, \tag{8}$$

where τ_y is the apparent yield stress of the material; K and n are the consistency and exponent of the power-law term in the flow curve; and $\zeta(\dot{\gamma})$ is a regularization factor given by

$$\zeta(\dot{\gamma}) = \begin{cases} 1 & \text{if } \tau \geq \tau_y \\ \frac{(\eta_0 - \eta_\infty)\dot{\gamma}}{\tau_y + K\dot{\gamma}^n} & \text{if } \tau < \tau_y \end{cases} \tag{9}$$

The dimensionless parameters associated with the flow curve are obtained by inputting $\dot{\gamma}_c$ in Equation (8), discarding the regularization factor, and dividing the resulting equation by the characteristic stress, τ_c. This procedure leads to

$$1 = \tau_y^* + K^* + \eta_\infty^*, \tag{10}$$

where

$$\tau_y^* = \frac{\tau_y}{\tau_c}, \quad K^* = \frac{K}{\tau_c}\left(\frac{U}{D}\right)^n, \quad \eta_\infty^* = \frac{\eta_\infty}{\eta_c} \tag{11}$$

with the characteristic viscosity defined as $\eta_c \equiv \tau_c D/U$. With this definition, differently from the Bingham number, the plastic number, τ_y^*, is a normalized quantity.

Assuming $\tau = \sigma_{dev}$, where σ_{dev} is the deviatoric part of the stress tensor, we are able to solve (8) for implicitly computing the deformation rate value, $\dot{\gamma}$, which is used for determining the viscosity through

$$\eta = \frac{\sigma_{dev}}{\dot{\gamma}}. \qquad (12)$$

Once the viscosity is obtained, the structure parameter is computed as

$$\lambda = \frac{\ln\left(\frac{\eta}{\eta_\infty}\right)}{\ln\left(\frac{\eta_0}{\eta_\infty}\right)}. \qquad (13)$$

It is worth noticing that the mathematical model used in this current work is similar that one proposed by Saramito [53] for dealing with yield-stress materials.

The equations presented in this section are solved in the context of a finite difference scheme. In particular, the continuity and momentum Equations (1) and (2) are solved via the projection method while the constitutive Equation (3) is computed via an explicit time discretization. The non-linear Equation (8) for variable $\dot{\gamma}$ is solved using a hybrid version of the Newton-bissection scheme.

3. Flow Problem Description and Moving Interface Dynamics

The flow problem addressed in this work consists of the impact of a fluid drop with diameter D_0 on a vertical wall, as illustrated in Figure 1, where one can also find (in the right) the computational domain filled with the mesh employed after the mesh study presented below.

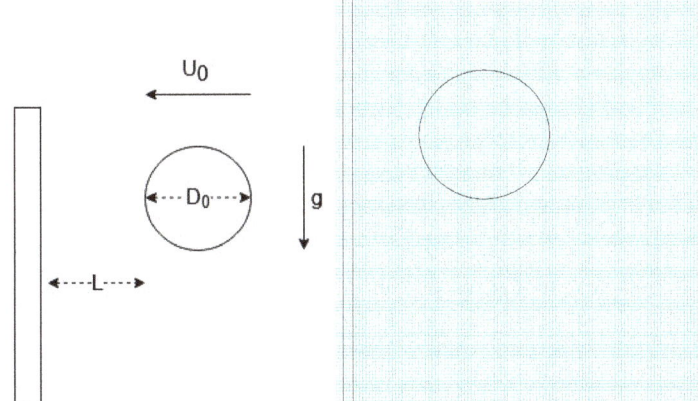

Figure 1. Schematic (**left**) and computational domain (**right**) of the problem considered.

We have imposed no-slip and no-penetration boundary condition, i.e., $\mathbf{u} = \mathbf{0}$ at the rigid wall. For the drop interface, the boundary conditions for the free surface are given by the following equations:

$$\mathbf{n} \cdot \sigma \cdot \mathbf{n}^T = 0, \qquad (14)$$

$$\mathbf{m} \cdot \sigma \cdot \mathbf{n}^T = 0, \qquad (15)$$

where $\mathbf{n}$ and $\mathbf{m}$ are the unit normal and tangent vectors, respectively. Equation (14) reflects the assumption of negligible surface tension effects.

To model the moving free surface interface, we have adopted a Front-Tracking method combined with the Marker-And-Cell scheme. In summary, in this framework, the interface is explicitly represented by massless marker particles that are advected by the flow as a passive quantity, i.e., solving the equation

$$\frac{d\mathbf{x}}{dt} = \mathbf{u}. \tag{16}$$

More details concerning the combination of Marker-And-Cell method and Front-Tracking scheme can be found in the previous works [45,50].

4. Results

This section is divided into two parts. The first one is dedicated to the verification of the code while in the second part of the results we investigate the role of the different dimensionless numbers of the problem, by means of parametric studies.

For the verification part, we performed a mesh refinement study using mechanical properties of real fluids adopting the material data of the Carbopol and Kaolin water solutions obtained from the literature.

In the context of the parametric study, we have disposed the investigation in three groups of cases. In the first group, gravitational effects are captured by analyzing the role of the Froude number for a high and a low value of the Weissenberg number keeping low levels of inertia. The second group analyses the role of elasticity for a high and a low value of the Froude number and large inertial effects. Finally, the third group explores the plastic number in the scenario where the Froude number is high, and inertia is low. In all simulations of the parametric study we have fixed the value of the dimensionless unstructured viscosity as $\eta^*_\infty = 0.2$. Drop evolution is presented as a function of the dimensionless time, $t^* = tU/D_0$.

From now on, the dimensionless numbers are presented without the $*$ symbol.

4.1. Code Verification

For the mesh refinement study, we have selected three meshes, for instance, M0, with 40 cells in the diameter of the drop ($\delta x^* = \delta y^* = 0.025$), M1, with 80 cells in the diameter of the drop ($\delta x^* = \delta y^* = 0.0125$) and M2, with 160 cells in the diameter of the drop ($\delta x^* = \delta y^* = 0.00625$).

In this Section we have considered two real yield stress fluids commonly used in the literature, e.g., Carbopol and Kaolin. The data of Carbopol and Kaolin fluids were respectively obtained from [54] and [39]. We set the following values for the dimensionless parameters: (i) Carbopol with $Re = 19.71$, $Wi = 1.21$, $\tau_y = 0.41$, $K = 0.58$ and (ii) Kaolin with $Re = 6.75$, $Wi = 0.00652$, $\tau_y = 0.21$, $K = 0.78$. In the analysis of both fluids, we kept the same Froude number and unstructured viscosity as $Fr = 2.26$ and $\eta_\infty = 0.01$.

Results for the evolution of the free surface shape of the drops considering M0, M1 and M2 for the two fluids at different times are shown in Figure 2 (see video0.mp4 for the complete evolution). Additional results for the mesh refinement study are presented in Figure 3. In this figure, we have plotted the maximum spreading length in the y-direction relative to the initial drop diameter as a function of time for Carbopol and Kaolin. From Figures 2 and 3, we can confirm the adequate convergence of the free surface with a mesh refinement. In particular, the results for M1 are slightly different those from M2; therefore, in order to save CPU time, we have adopted M1 for the remaining simulations in this work.

From a physical point-of-view, we can conclude from Figure 3a that the elastic effects in the Carbopol solution are more evident since the drop presents oscillations after the vertical impact. On the other hand, after achieving the maximum size in y-direction, the drop of Kaolin does not retract, but stays stuck on the vertical wall, in a typical plastic deformation. These results are explained by the discrepancy in the Weissenberg number between the solutions

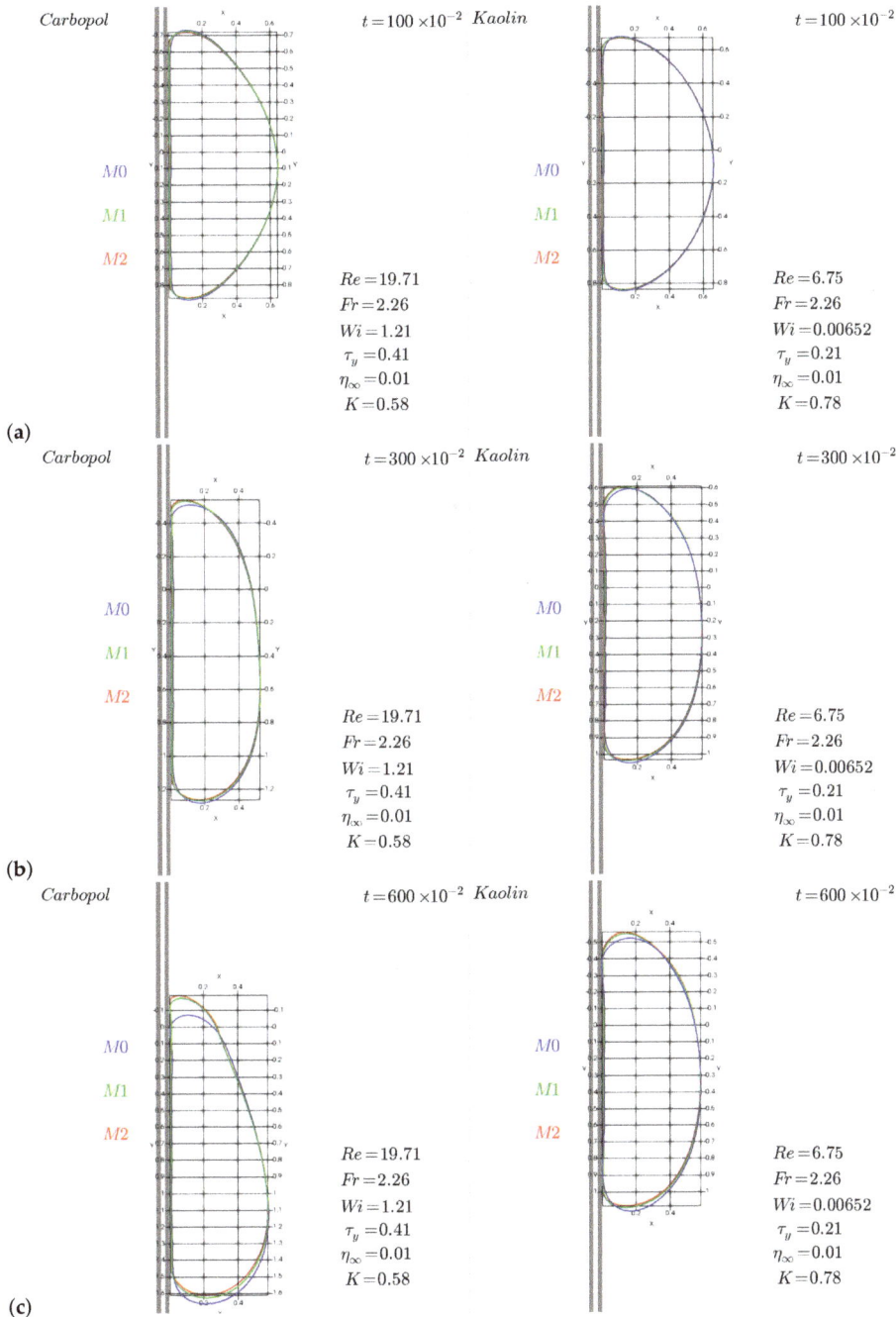

Figure 2. Collision of drops on a vertical wall for Carbopol (**left**) and Kaolin (**right**) materials using three different meshes (M1, M2 and M3): (**a**) Dimensionless time, $t = 1$; (**b**) Dimensionless time, $t = 3$ and (**c**) Dimensionless time, $t = 6$.

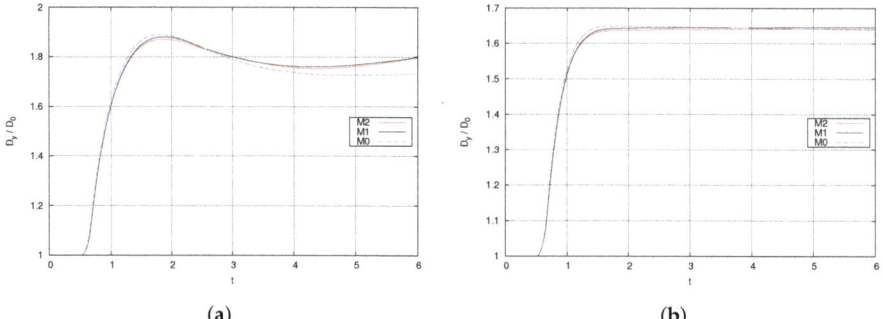

(a)   (b)

Figure 3. Vertical diameter as function of the time: Carbopol (**a**) and Kaolin (**b**).

4.2. Group 1

In this first group, we maintained fixed the values of the Reynolds number, $Re = 3.0$, and the plastic number, $\tau_y = 0.4$, while the Froude number values examined were $Fr = \{1.69; 2.26; 4.51; 9.03\}$. Figures 4–7 show the evolution of the collision of elasto-viscoplastic drops on a vertical plane for this group. In order to better illustrate the interface dynamics of these simulations, we have included videos related to these figures as supplementary materials (see video1.mp4). The time values shown in these figures are $t = \{2; 4; 6; 34\}$. The left column corresponds to a very low Weissenberg number, $Wi = 0.01$, which can be considered to be a viscoplastic material since the elastic effects are negligible. In the right column, the value $Wi = 1$ corresponds to cases where the elastic nature of the material is significant.

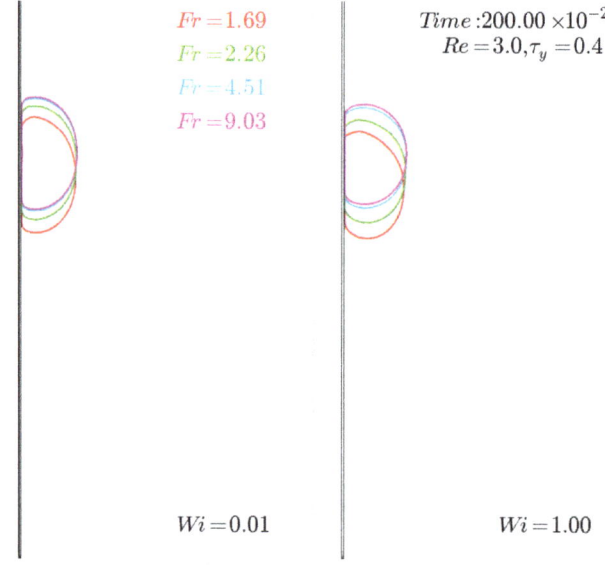

Figure 4. Collision of drops on a vertical wall for fixed values of the Reynolds, $Re = 3$, and plastic numbers, $\tau_y = 0.4$ and the following values of the Froude number, $Fr = \{1.69; 2.26; 4.51; 9.03\}$. (**left column**) $Wi = 0.01$; (**right column**) $Wi = 1.0$. Dimensionless time, $t = 2$.

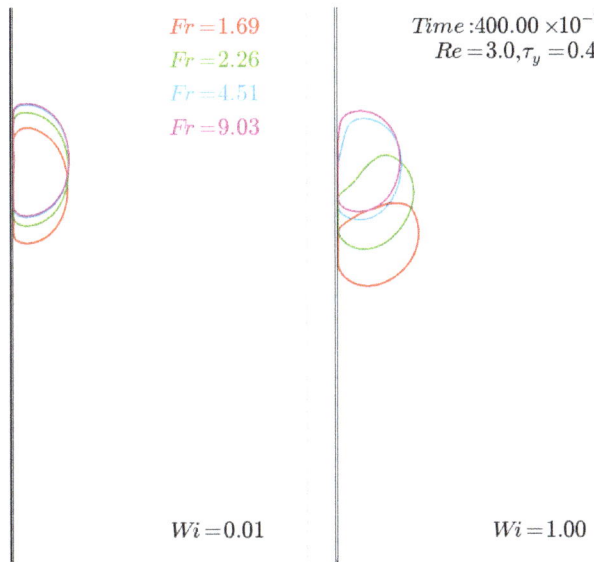

Figure 5. Collision of drops on a vertical wall for fixed values of the Reynolds, $Re = 3$, and plastic numbers, $\tau_y = 0.4$ and the following values of the Froude number, $Fr = \{1.69; 2.26; 4.51; 9.03\}$. (**left column**) $Wi = 0.01$; (**right column**) $Wi = 1.0$. Dimensionless time, $t = 4$.

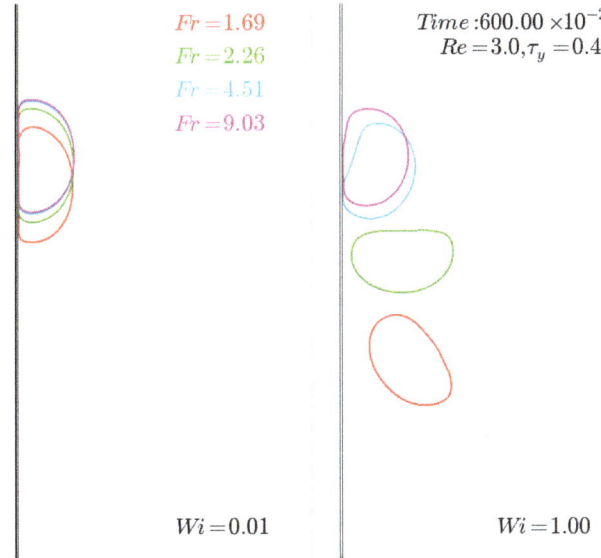

Figure 6. Collision of drops on a vertical wall for fixed values of the Reynolds, $Re = 3$, and plastic numbers, $\tau_y = 0.4$ and the following values of the Froude number, $Fr = \{1.69; 2.26; 4.51; 9.03\}$. (**left column**) $Wi = 0.01$; (**right column**) $Wi = 1.0$. Dimensionless time, $t = 6$.

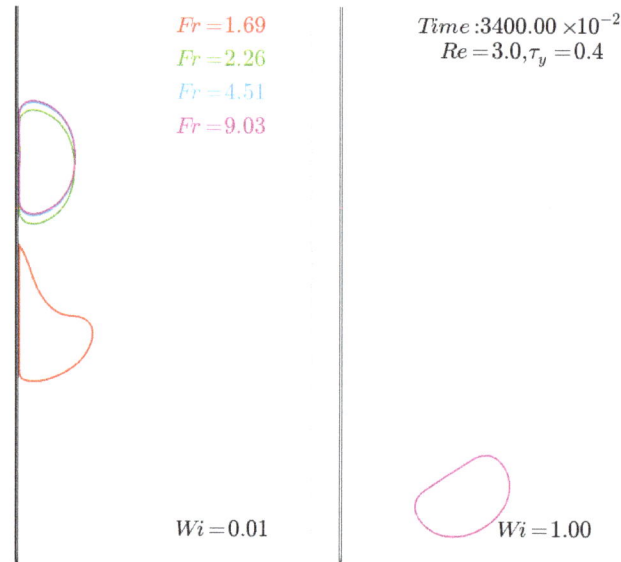

Figure 7. Collision of drops on a vertical wall for fixed values of the Reynolds, $Re = 3$, and plastic numbers, $\tau_y = 0.4$ and the following values of the Froude number, $Fr = \{1.69; 2.26; 4.51; 9.03\}$. (**left column**) $Wi = 0.01$; (**right column**) $Wi = 1.0$. Dimensionless time, $t = 34$.

A remarkable result is the connection between elasticity and the detachment of the drop from the solid surface. For $Wi = 0.01$ all the drops remained attached to the vertical wall. In these cases, the gravitational effects, which are more pronounced in the lowest Fr case, act to dislocate the drop downwards. We can observe that as the Froude number is decreased the drop goes through a higher dislocation, as expected. For the lowest Fr case, in addition to reaching the lowest position, the induced deformation is acting to inflate its bottom part, as illustrated by the $Fr = 1.69$-drop on the left column of Figure 7. This happens because the wall imposes a resistance to the fluid, which is in contact to the solid, while far from the wall, gravity acts more freely. Similar effects were found by Podgorski et al. [19] and Dimitrakopoulos & Higdon [18] when gravitational effects were much higher than surface tension ones. On the other hand, the high elasticity promoted by the $Wi = 1$ value of the Weissenberg number was able to induce a detachment from the wall for all the values of Froude number analyzed, including the highest Froude number case, $Fr = 9.03$. As expected, lower values of Fr anticipates the detachment of the drop. It is worth noticing that the elasticity of the drop plays the role in this case of the surface tension in Newtonian drops where capillary effects are present, i.e., the stored elastic energy during the spreading stage is able to induce receding and the detachment of the drop.

4.3. Group 2

In this second group of cases, we increased the value of the Reynolds number to $Re = 8.0$, and kept the plastic number in the same level, $\tau_y = 0.4$. The high and low values of the Froude number are $Fr = 4.52$ (left columns) and $Fr = 1.13$ (right columns). Figures 8–11 correspond to dimensionless time values given by $t = \{1; 3; 6; 20\}$. Videos related to these figures can be found in the supplementary materials (see video3.mp4). The values of the Weissenberg number shown in each frame are $Wi = \{0.01; 0.1; 0.5; 1\}$, ranging from a viscoplastic behavior to a reasonably elastic one.

The gravitational effects are very highlighted in this group of cases. In the first captured frame, $t = 1$, elastic effects did not have time to appear and the drops on the left and on the right are almost coincident. At this time, the only Froude number effect was to dislocate downwards the drops on the right. At the time $t = 3$, we notice a significant gravitational effect on the right column where the drop has slithered down the vertical wall. This kind of behavior was already found in [55] for the problem of a material deposited on an inclined plane. It is worth observing an important difference with respect to the previous group which is the highest Reynolds number value in this case. The additional energy that came from the kinetic energy source was able to induce a more stretched deformation that was amplified by gravity. Figure 10 reveals that the less elastic drops ($Wi = 0.01$ and $Wi = 0.10$) of the higher-Forude-number column (right) exhibited a "pinned tail", i.e., the top tip remains at fixed position. This condition is in contrast with respect to the drops with higher Wi, where an elastic retraction takes place. The left column shows the beginning of a detachment process of the highest Wi case that ends in Figure 11, where this drop begins to fall. For the low values of the Weissenberg number, the weak elastic effects were unable to detach the drops. In all cases, the high initial kinetic energy was mostly dissipated in the impact and spreading. The weak gravitational effects for the low Fr cases showed more symmetric drops as final configurations.

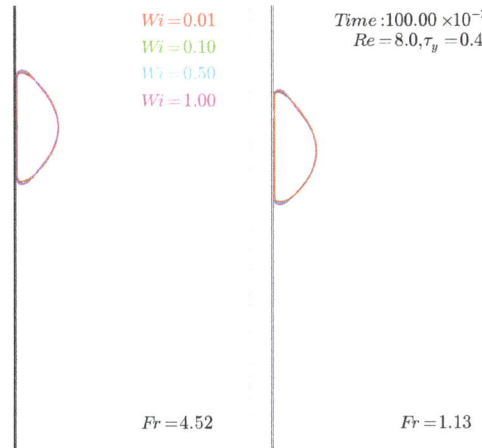

Figure 8. Collision of drops on a vertical wall for fixed values of the Reynolds, $Re = 8$, and plastic numbers, $\tau_y = 0.4$ and the following values of the Weissenberg number, $Wi = \{0.01; 0.1; 0.5; 1\}$. (**left column**) $Fr = 4.52$; (**right column**) $Fr = 1.13$. Dimensionless time, $t = 1$.

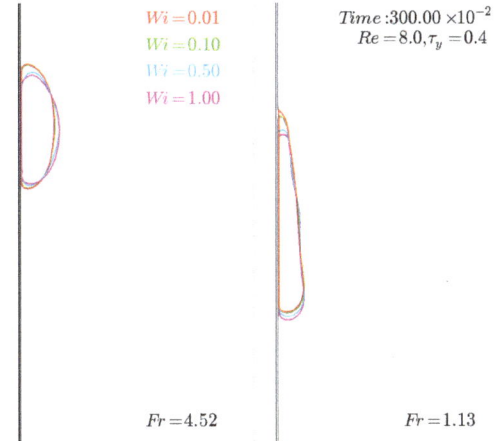

Figure 9. Collision of drops on a vertical wall for fixed values of the Reynolds, $Re = 8$, and plastic numbers, $\tau_y = 0.4$ and the following values of the Weissenberg number, $Wi = \{0.01; 0.1; 0.5; 1\}$. (**left column**) $Fr = 4.52$; (**right column**) $Fr = 1.13$. Dimensionless time, $t = 3$.

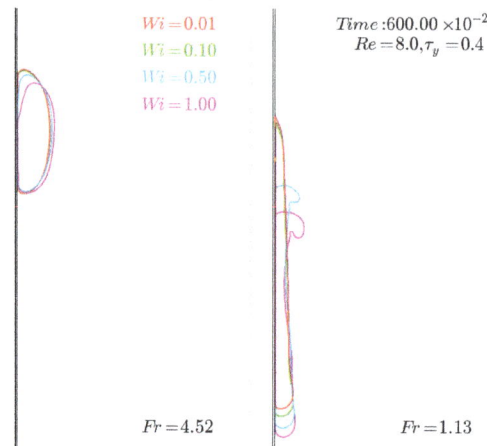

Figure 10. Collision of drops on a vertical wall for fixed values of the Reynolds, $Re = 8$, and plastic numbers, $\tau_y = 0.4$ and the following values of the Weissenberg number, $Wi = \{0.01; 0.1; 0.5; 1\}$. (**left column**) $Fr = 4.52$; (**right column**) $Fr = 1.13$. Dimensionless time, $t = 6$.

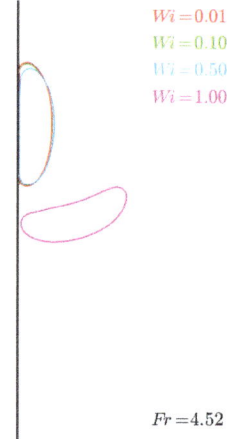

Figure 11. Collision of drops on a vertical wall for fixed values of the Reynolds, $Re = 8$, and plastic numbers, $\tau_y = 0.4$ and the following values of the Weissenberg number, $Wi = \{0.01; 0.1; 0.5; 1\}$. (**left column**) $Fr = 4.52$; (**right column**) $Fr = 1.13$. Dimensionless time, $t = 20$.

4.4. Group 3

In this third group of cases, inertia and gravitational effects are low, with $Re = 2.0$ and $Fr = 9.03$. Three columns are shown in Figures 12–14, corresponding to different levels of the plastic number (see video2.mp4 in supplementary materials). From left to right, the values of the plastic number are $\tau_y = 0.1$, $\tau_y = 0.2$, $\tau_y = 0.4$. The values of the Weissenberg number shown in each frame are $Wi = \{0.01; 0.1; 0.5; 1.0; 2.0\}$.

After impact, as shown by Figure 12 for $t = 1$, there is no significant difference among the three columns. We can notice a tinny variation with respect to the elasticity of the fluid. Higher values of Wi lead to more deformed drops. This is expected because significant elastic effects in the case of viscoelastic fluids are connected with lower rigidity, what in turn favors deformation. As time evolves to $t = 4$ we notice a rebound of the more elastic drop in the highest plastic number, Figure 13 right column. We can see that this bouncing occurs in the horizontal direction, revealing that gravity did not play a role until this moment, due to the low Froude number. However, differently from an impact on a horizontal plane, we have no kicking after a first rebound, since the gravitational force does not induce the drop to return to the plane, as in the collision on a floor. In this regard, we see in Figure 14 that the drops of high Weissenberg numbers detached from the vertical wall and have fallen down. The plastic number effects for this set of cases were not significant.

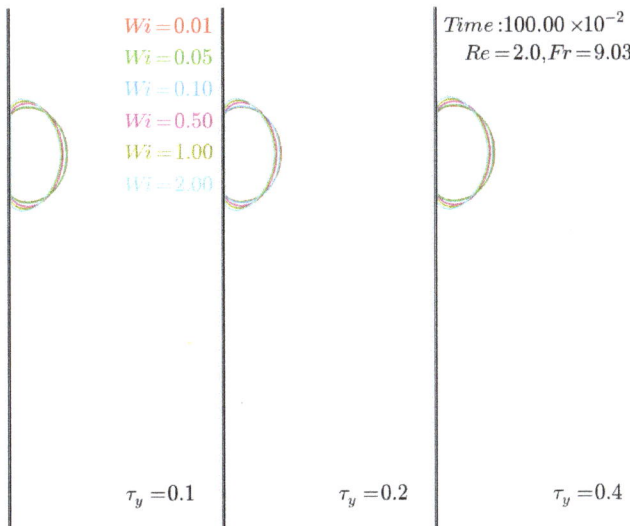

Figure 12. Collision of drops on a vertical wall for fixed values of the Reynolds, $Re = 2$, and Froude numbers, $Fr = 9.03$ and the following values of the Weissenberg number, $Wi = \{0.01; 0.1; 0.5; 1; 2\}$. (**left column**) $\tau_y = 0.1$; (**right column**) $\tau_y = 0.2$; (**right column**) $\tau_y = 0.4$. Dimensionless time, $t = 1$.

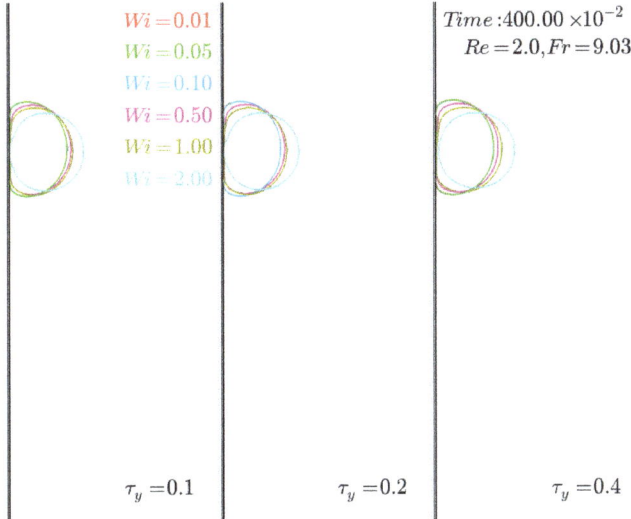

Figure 13. Collision of drops on a vertical wall for fixed values of the Reynolds, $Re = 8$, and plastic numbers, $\tau_y = 0.4$ and the following values of the Weissenberg number, $Wi = \{0.01; 0.1; 0.5; 1\}$. (**left column**) $Fr = 4.52$; (**right column**) $Fr = 1.13$. Dimensionless time, $t = 4$.

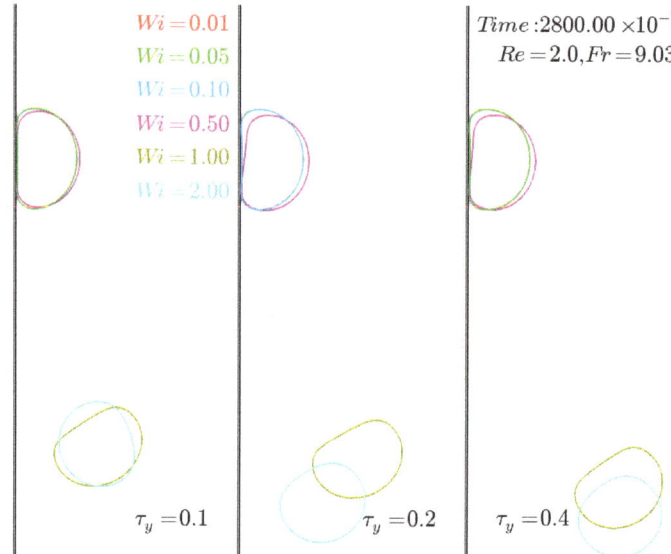

Figure 14. Collision of drops on a vertical wall for fixed values of the Reynolds, $Re = 8$, and plastic numbers, $\tau_y = 0.4$ and the following values of the Weissenberg number, $Wi = \{0.01; 0.1; 0.5; 1\}$. (**left column**) $Fr = 4.52$; (**right column**) $Fr = 1.13$. Dimensionless time, $t = 28$.

5. Summary and Conclusions

In the present work we investigated the collision of elasto-viscoplastic drops on a vertical plane. Special emphasis is given on the gravitational effects of this problem. The combination of the dimensionless numbers that govern the problem, namely the Reynolds, Weissenberg, plastic, and Froude numbers, were able to induce a rich variety of outputs. Gravitational effects anticipate the detachment of more elastic drops from the wall and induce large deformations in less elastic drops in the cases of low impact velocities. In addition, gravity favors sliding and slithering of the drops that collide with high impact velocities. Differently from horizontal planes, in vertical ones there is no kicking and, therefore, bouncing refrains the drop from being in contact with the solid, what can affect significantly real applications.

Author Contributions: Methodology, F.P.M. and R.L.T.; software, C.M.O. and F.P.M.; formal analysis, R.L.T.; investigation, F.P.M. and R.L.T.; resources, C.M.O.; writing—original draft preparation, R.L.T.; writing—review and editing, C.M.O. All authors have read and agreed to the published version of the manuscript.

Funding: This research was partially supported by funds of CNPq (National Council for Scientific and Technological Development) grants 305383/2019-1 and, FAPESP (2013/07375-0) and CAPES.

Conflicts of Interest: The authors declare no conflict of interest.

Abbreviations

The following abbreviations are used in this manuscript:

PEG   Polyethylene Glycol
PEO   Polyethylene Oxide

References

1. Josserand, C.; Thoroddsen, S.T. Drop impact on a solid surface. *Ann. Rev. Fluid Mech.* **2016**, *48*, 365–391. [CrossRef]
2. Philippi, J.; Lagrée, P.-Y.; Antkowiak, A. Drop impact on a solid surface: Short-time self-similarity. *J. Fluid Mech.* **2016**, *795*, 96–135. [CrossRef]
3. Pasandideh-Fard, M.; Qiao, Y.M.; Chandra, S.; Mostaghimi, J. Capillary effects during droplet impact on a solid surface. *Phys. Fluids* **1996**, *8*, 650–659. [CrossRef]
4. Yarin, A.L. Drop impact dynamics: Splashing, spreading, receding, bouncing. *Ann. Rev. Fluid Mech.* **2006**, *38*, 159–192. [CrossRef]
5. De Gennes, P.G. Wetting: Statics and dynamics. *Rev. Mod. Phys.* **1985**, *57*, 827–863. [CrossRef]
6. Sikalo, S.; Wilhelm, H.D.; Roisman, I.V.; Jakirlic, S.; Tropea, C. Dynamic contact angle of spreading droplets: Experiments and simulations. *Phys. Fluids* **2005**, *17*, 062103. [CrossRef]
7. Bayer, I.S.; Megaridis, C.M. Contact angle dynamics in droplets impacting on flat surfaces with different wetting characteristics. *J. Fluid Mech.* **2006**, *558*, 415–449. [CrossRef]
8. Vadillo, D.; Soucemarianadin, A.; Delattre, C.; Roux, D. Dynamic contact angle effects onto the maximum drop impact spreading on solid surfaces. *Phys. Fluids* **2009**, *21*, 122002. [CrossRef]
9. Antonini, C.; Villa, F.; Marengo, M. Oblique impacts of water drops onto hydrophobic and superhydrophobic surfaces: Outcomes, timing, and rebound maps. *Exp. Fluids* **2014**, *55*, 1–9. [CrossRef]
10. Yeong, Y.H.; Burton, J.; Loth, E. Drop impact and rebound dynamics on an inclined superhydrophobic surface. *Langmuir* **2014**, *30*, 12027–12038. [CrossRef]
11. Khojasteha, D.; Kazeroonib, M.; Salariana, S.; Kamalia, R. Droplet impact on superhydrophobic surfaces: A review of recent developments. *J. Ind. Eng. Chem.* **2016**, *42*, 1–14. [CrossRef]
12. Antonini, C.; Jung, S.; Wetzel, A.; Heer, E.; Schoch, P.; Moqaddam, A.M.; Chikatamarla, S.S.; Karlin, I.; Marengo, M.; Poulikakos, D. Contactless prompt tumbling rebound of drops from a sublimating slope. *Phys. Rev. Fluids* **2016**, *1*, 013903. [CrossRef]
13. Bird, J.C.; Dhiman, R.; Varanasi, K.K. Reducing the contact time of a bouncing drop. *Nature* **2013**, *503*, 385–397. [CrossRef] [PubMed]
14. Liu, Y.; Andrew, M.; Li, J.; Yeomans, J.M.; Wang, Z. Symmetry breaking in drop bouncing on curved surfaces. *Nat. Comm* **2015**, *6*, 10034. [CrossRef]
15. Xie, J.; Xu, J.; Shang, W.; Zhang, K. Mode selection between sliding and rolling for droplet on inclined surface: Effect of surface wettability. *Int. J. Heat Mass Transf.* **2018**, *122*, 45–58. [CrossRef]
16. Dussan, V.E.B.; Chow, R.T.-P. On the ability of drops or bubbles to stick to non–horizontal surfaces of solids. *J. Fluid Mech.* **1983**, *137*, 1–29. [CrossRef]
17. Dussan, V.E.B. On the ability of drops or bubbles to stick to non-horizontal surfaces of solids. Part 2. Small drops or bubbles having contact angles of arbitrary size. *J. Fluid Mech.* **1985**, *151*, 1–20. [CrossRef]
18. Dimitrakopoulos, P.; Higdon, J.J.L. On the gravitational displacement of three-dimensional fluid droplets from inclined solid surfaces. *J. Fluid Mech.* **1999**, *395*, 181–209. [CrossRef]
19. Podgorski, T.; Flesselles, J.M.; Limat, L. Corners, Cusps, and Pearls in Running Drops. *Phys. Rev. Lett.* **2001**, *87*, 036102. [CrossRef]
20. Milinazzo, F.; Shinbrot, M. A numerically study of a drop on a vertical wall. *J. Colloid Interface Sci.* **1988**, *121*, 254–264. [CrossRef]
21. Schwartz, L.W.; Roux, D.; Cooper-White, J.J. On the shapes of droplets that are sliding on a vertical wall. *Physica D* **2005**, *209*, 236–244. [CrossRef]
22. Wirth, W.; Storp, S.; Jacobsen, W. Mechanisms controlling leaf retention of agricultural spray solutions. *Pestic. Sci.* **1991**, *33*, 411–421. [CrossRef]
23. Glass, C.R.; Walters, K.F.A.; Gaskell, P.H.; Lee, Y.C.; Thompson, H.M.; Emerson, D.R.; Gu, X.J. Recent advances in computational fluid dynamics relevant to the modelling of pesticide flow on leaf surfaces. *Pest Manag. Sci.* **2010**, *66*, 2–9. [CrossRef] [PubMed]
24. Koch, K.; Bhushan, B.; Bathlott, W. Multifunctional surface structures of plants: An inspiration for biomimetics. *Prog. Matter Sci* **2009**, *54*, 137–178. [CrossRef]
25. Veremieiev, S.; Brown, A.; Gaskell, P.H.; Glass, C.R.; Kapur, N.; Thompson, H.M. Modelling the flow of droplets of bio-pesticide on foliage. *Interf. Phenom. Heat Trans.* **2014**, *2*, 1–14. [CrossRef]

26. Zhao, L.M.; Zhang, Y.M.; Zhao, H.Z.; Fang, J.; Qin, J. Numerical case studies of vertical wall fire protection using water spray. *Case Stud. Therm. Eng.* **2014**, *4*, 129–135. [CrossRef]
27. Zhuang, D.; Hu, H.; Ding, G.; Xi, G.; Han, W. Numerical model for liquid droplet motion on vertical plain–fin surface. *HVAC R Res.* **2014**, *20*, 332–343. [CrossRef]
28. Duy, V.N.; Vu, T.V. A numerical study of a liquid drop solidifying on a vertical cold wall. *Int. J. Heat Mass Transf.* **2018**, *127*, 302–312. [CrossRef]
29. Demidovich, A.V.; Kropotova, S.S.; Piskunov, M.V.; Shlegel, N.E.; Vysokomornaya, O.V. The impact of single- and multicomponent liquid drops on a heated wall: Child droplets. *Appl. Sci.* **2020**, *10*, 942. [CrossRef]
30. Xu, H.; Clarke, A.; Rothstein, J.P.; Poole, R.J. Viscoelastic drops moving on hydrophilic and superhydrophobic surfaces. *J. Colloid Int. Sci.* **2018**, *513*, 53–61. [CrossRef]
31. Jalaal, M.; Balmforth, N.J.; Stoeber, B. Slip of spreading viscoplastic droplets. *Langmuir* **2015**, *31*, 20171–20175. [CrossRef] [PubMed]
32. Blackwell, B.C.; Deetjen, M.E.; Gaudio, J.E.; Ewoldt, R.H. Sticking and splashing in yield-stress fluid drop impacts on coated surfaces. *Phys. Fluids* **2015**, *27*, 043101. [CrossRef]
33. Sen, S.; Morales, A.G.; Ewoldt, R.H. Viscoplastic drop impact on thin films. *J. Fluid Mech.* **2020**, *891*, 1–24, . [CrossRef]
34. Smith, M.; Bertola, V. Effect of polymer additives on the wetting of impacting droplets. *Phys. Rev. Lett.* **2010**, *104*, 154502. [CrossRef] [PubMed]
35. Bergeron, V.; Bonn, D.; Martin, J.Y.; Vovelle, L. Controlling droplet deposition with polymer additives. *Nature* **2000**, *405*, 772–775. [CrossRef] [PubMed]
36. Bartolo, D.; Boudaoud, A.; Narcy, G.; Bonn, D. Dynamics of non-newtonian droplets. *Phys. Rev. Lett.* **2007**, *99*, 175502. [CrossRef] [PubMed]
37. Bertola, V. The impact of viscoplastic drops on a heated surface in the Leidenfrost regime. *Soft Matter* **2012**, *12*, 7624–7631.
38. Rozhkov, A.; Prunet-Foch, B.; Vignes-Adler, M. Impact of drops of polymer solutions on small targets. *Phys. Fluids* **2003**, *15*, 2006. [CrossRef]
39. Luu, L.H.; Forterre, Y. Drop impact of yield-stress fluids. *J. Fluid Mech.* **2009**, *632*, 301–327. [CrossRef]
40. Nigen, S. Experimental investigation of the impact of an (apparent) yield-stress material. *Atom. Spray* **2005**, *15*, 103–117. [CrossRef]
41. German, G.; Bertola, V. Impact of shear-thinning and yield-stress drops on solid substrates. *J. Phys. Condens. Matter* **2009**, *21*, 375111. [CrossRef] [PubMed]
42. Saidi, A.; Martin, C.; Magnin, A. nfluence of yield stress on the fluid droplet impact control. *J. Non-Newton. Fluid Mech.* **2010**, *165*, 596–606. [CrossRef]
43. Kim, E.; Baek, J. Numerical study of the parameters governing the impact dynamics of yield-stress fluid droplets on a solid surface. *J. Non-Newton. Fluid Mech.* **2012**, *173–174*, 62–71. [CrossRef]
44. Fang, J.; Owens, R.G.; Tacher, L.; Parriaux, A. A numerical study of the SPH method for simulating transient viscoelastic free surface flows. *J. Non-Newton. Fluid Mech.* **2006**, *139*, 68–84. [CrossRef]
45. Oishi, C.M.; Martins, F.P.; Tome, M.F.; Alves, M.A. Numerical simulation of drop impact and jet buckling problems using the extended pom-pom model. *J. Non-Newton. Fluid Mech.* **2012**, *169–170*, 91–103. [CrossRef]
46. Noroozi, S.; Tavangar, S.; Hashemabadi, S.H. CFD simulation of wall impingement of tear shape viscoplastic drops utilizing OpenFOAM. *Appl. Rheol.* **2013**, *23*, 55519.
47. Izbassarov, D.; Muradoglu, M. Effects of viscoelasticity on drop impact and spreading on a solid surface. *Phys. Rev. Fluids* **2016**, *1*, 023302. [CrossRef]
48. Wang, Y.; Do-Quang, M.; Amberg, G. Impact of viscoelastic droplets. *J. Non-Newton. Fluid Mech.* **2017**, *243*, 38–46. [CrossRef]
49. Oishi, C.M.; Thompson, R.L.; Martins, F.P. Normal and oblique drop impact of yield-stress fluids with thixotropic effects. *J. Fluid Mech.* **2019**, *876*, 642–679. [CrossRef]
50. Oishi, C.M.; Thompson, R.L.; Martins, F.P. Impact of capillary drops of complex fluids on a solid surface. *Phys. Fluids* **2019**, *31*, 123109. [CrossRef]
51. Thompson, R.L.; Soares, E.J. Viscoplastic dimensionless numbers. *J. Non-Newton. Fluid Mech.* **2016**, *238*, 57–64. [CrossRef]

52. Oishi, C.M.; Thompson, R.L.; Martins, F.P. Transient motions of elasto-viscoplastic thixotropic materials subjected to an imposed stress field and to stress-based free-surface boundary conditions. *Int. J. Eng. Sci.* **2016**, *109*, 165–201. [CrossRef]
53. Saramito, P. A new elastoviscoplastic model based on the Herschel–Bulkley viscoplastic model. *J. Non-Newton. Fluid Mech.* **2009**, *158*, 154–161. [CrossRef]
54. Jalaal, M.; Kemper, D.; Lohse, D. Viscoplastic water entry. *J. Fluid Mech.* **2019**, *864*, 596–613. [CrossRef]
55. Oishi, C.M.; Martins, F.P.; Thompson, R.L. The "avalanche effect" of an elasto-viscoplastic thixotropic material on an inclined plane. *J. Non-Newton. Fluid Mech.* **2017**, *247*, 165–201. [CrossRef]

 © 2020 by the authors. Licensee MDPI, Basel, Switzerland. This article is an open access article distributed under the terms and conditions of the Creative Commons Attribution (CC BY) license (http://creativecommons.org/licenses/by/4.0/).

Article

Effects of Polypropylene Fibers and Measurement Methods on the Yield Stress of Grouts for the Consolidation of Heritage Masonry Walls

Luis G. Baltazar [1,*], **Fernando M. A. Henriques** [1] **and Maria Teresa Cidade** [2]

[1] Departamento de Engenharia Civil, Faculdade de Ciências e Tecnologia, Universidade NOVA de Lisboa, 2829-516 Caparica, Portugal; fh@fct.unl.pt
[2] Departamento de Ciência dos Materiais e CENIMAT/I3N, Faculdade de Ciências e Tecnologia, Universidade NOVA de Lisboa, 2829-516 Caparica, Portugal; mtc@fct.unl.pt
* Correspondence: luis.baltazar@fct.unl.pt; Tel.: +35-121-294-8300

Received: 27 February 2020; Accepted: 16 April 2020; Published: 20 April 2020

Abstract: The injection of grouts is a consolidation technique suitable for overcoming the structural deterioration of old stone masonry walls. Grouting operations involve introducing a suspension (grout) into a masonry core with the aim of improving the load capacity of the wall, as well as reducing its brittle mechanisms. The yield stress of injection grouts will affect the injection pressure and their flow inside the masonry. However, the determination of some rheological properties such as yield stress in hydraulic grout is challenging, due to the combined effects of hydration reactions and interactions between the particles present in the suspension. In this study, the determination of the yield stress of natural hydraulic lime-based grouts with polypropylene fibers was carried out. The changes in yield stress with time, fibers content and hydration were evaluated by two measurement methods using a rotational rheometer. Additionally, the static and dynamic yield stress as well as the critical shear–strain rate were determined, which provided useful information on the grout design in order to achieve successful grouting operations.

Keywords: yield stress; grout; polypropylene fiber; masonry; consolidation; rheology

1. Introduction

Ordinary or historical buildings in most european cities until the mid-20th century were built with stone masonry walls. Stone masonry is characterized by a certain vulnerability, mainly due to its irregular morphology and the presence of voids and loose adhesive material, which compromises its structural integrity. The injection of grouts (or grouting) is a consolidation technique to overcome masonry structural deterioration. Indeed, grout injections have been revealed to be an effective method to improve the load capacity of the walls, as well as reducing the brittle mechanisms [1]. Grouting operations involve introducing a suspension (grout) into the masonry core. For proper strengthening with this grouting technique, the compatibility of the new materials with the existing ones as well as good injectability are required. The grout's injectability appears to be one of the most crucial characteristics in the grouting's performance. In this context, rheology is used as a tool in the design and quality control of the injection grout [2,3]. Cementitious-based grouts are known for having a complex rheology [4,5], because interaction between the binder particles and grout hydration generates a microstructure that leads to a yield stress that is not constant over time. The yield stress of a grout will affect the relationship between the injection pressure and flow and, therefore, will set the distance that the grout can penetrate in the masonry inner core.

In the literature, some controversy about the existence of yield stress can be found. Some authors say that yield stress is not real and only seems to be present due to limitations in the rheological measurements [4]. On the other hand, others allege that, despite the fact that yield stress does not exist, it makes sense to consider it for practical issues [5,6]. Moreover, several studies show that yield stress in fact exists and, when the suspension is thixotropic, the yield stress strongly depends on a build-up and break-down of the material [7–9]. According to Barnes [10], the concept of yield stress has been proved and is still useful in a wide range of applications once it is properly delineated. Another study [11] that dealt with cementitious suspensions defined the yield stress based on two values, which are the static yield stress and dynamic yield stress. The static yield stress is associated with the yield after the suspension leaves the rest, while the dynamic yield stress is linked with the stress required so that the flow continues to occur. However, the reliability in determining yield stress depends on the rheological models used and several factors involved in the measurement apparatus—such as the gap between shearing surfaces, type of geometry and slippage of geometry at low shear–strain rates—which cause the determination of yield stress to be even more complicated [12]. More recently, yield stress has been pointed out as key parameter in the characterization of the pumpability of suspensions like drilling muds [13]. Notwithstanding the relevance of yield stress, no standard procedure is available to determine this parameter; very often, yield stress is used as a single value without considering certain phenomena, such as the thixotropy, the range of the applied shear–strain rate and the hydration of the binder. This work intends to contribute to filling this knowledge gap.

Natural hydraulic lime (NHL) is the most commonly used material for heritage masonry consolidation since it is a binder that provides adequate chemical, physical and mechanical compatibility with the original materials present in the masonry [14]. Nevertheless, one of NHL's drawbacks is cracking due to drying shrinkage [15]. However, this issue can be overcome by adding fibers [16,17]. Fibers are additions that are frequently incorporated into cementitious materials aiming to increase their mechanical strength and reduce cracking [18,19]. There are various types of fibers that can be used in cementitious materials: natural, metallic, ceramic and polymeric [20]. They may be categorized according to the raw material and they can also be characterized in terms of their length, shape and diameter [17].

Polypropylene (PP) fiber is the most common type of fiber used in cementitious materials, but there are still a very limited number of studies on the effects of PP fiber on lime-based materials [21,22]. For instance, Chan and Bindiganavile [23,24] used it in NHL mortars with a dosage of up to 0.5% by volume. Moreover, Barbero-Barrera and Medina [25] analyzed the effect of PP fibers on graphite-natural hydraulic lime pastes. Besides the improvement of mechanical properties that PP fibers can provide, the rheological properties, especially the yield stress, of grouts containing PP fibers should also be examined for an accurate injection. Within this scope, this work aimed to determine both the static and dynamic yield stress of NHL grouts, including the contribution of thixotropy, time period and PP fibers content, by means of two measurement methods using a rotational rheometer. Additionally, several correlations between yield stress, time and critical shear–strain rate (marking the transition from dynamic to static yield stress) were developed.

2. Materials and Methods

2.1. Materials

The experimental program was conducted using a type of NHL, namely NHL5, produced according to the european Standard eN459-1:2010 [26]. NHL was chosen as the binder, since it is a hydraulic binder that has chemical and physical properties close to those of pre-existing materials in heritage masonries and can set both in dry and wet conditions [11,27]. The physical and chemical properties of NHL are listed in Table 1. The grain size distribution is represented in Figure 1.

A type of polypropylene (PP) fibers with a density of 0.9 g/cm^3 was used. The average diameter and length of the PP fibers were 0.05 and 12 mm, respectively. To improve grout's injectability, especially at a low water/binder ratio, superplasticizers are frequently used [28,29]. In this study, a commercially available polycarboxylate powder superplasticizer (i.e., high range water reducer) conforming to ASTM C494-05 [30] was used. It had a specific gravity, pH, and chloride content of 0.6, 4.2 and <0.1%, respectively.

Table 1. Chemical composition and physical property of natural hydraulic lime (NHL5) (provided by the manufacturer).

Parameter	Value
Compression resistance at 28 days	5.5 MPa
Setting time	Start: 2 h end: 6 h
Density	2.73 g/cm^3
Specific surface area B.e.T.	480 m^2/kg
Al_2O_3	2.00%
CaO	85.00%
Fe_2O_3	2.00%
MgO	1.00%
MnO	0.03%
SiO_2	8.00%
SiC	0.01%
SO_3	1.00%
SrO	0.05%
K_2O	0.70%

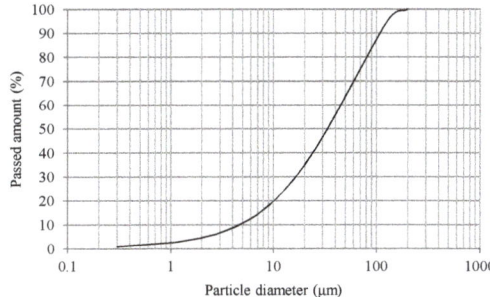

Figure 1. Particle size distribution of NHL.

2.2. Mixture Proportions and Procedure

In this study, all the grouts were based on a water-to-binder (w/b) ratio of 0.4 and the dosage of superplasticizer was set at 0.2% of the mass of NHL. Both the w/b ratio and superplasticizer dosage were chosen according to their typical usage in field applications [31]. The volume fractions of PP fibers by NHL mass were selected as 0%, 0.03% and 0.1%. The mix proportions of grouts are listed in Table 2.

Table 2. Proportions of grouts used in this study.

Notation	Water/Binder (-)	Superplasticizer (wt%)	PP Fiber (vol%)
Ref	0.4	0.2	0
0.03% PP	0.4	0.2	0.03
0.1% PP	0.4	0.2	0.1

All the grouts were prepared in laboratory in batches of 300 mL and mixed using a high shear mixer equipped with a helicoidal blade. The NHL and superplasticizer were first added into the mixer and dry-mixed at a low speed for 2 min, and the PP fibers were gradually introduced at the same time. Afterwards, water was added into the mixer and mixed at a high speed for 3 min. The grouts were prepared at an ambient temperature of 20 ± 2 °C and a relative humidity of 60% ± 5%. Following the completion of mixing, the rheological measurements were performed as described below.

2.3. Experimental Procedures

The rheological measurements were performed with a Bohlin Gemini HR^{nano} rotational rheometer (Malvern Instruments). Parallel-plate geometry was used to perform all the measurements. The diameter of the geometry was 40 mm and the gap was 2 mm. The surface roughness of the upper plate was modified by means of an emery paper (grid 120) to minimize slippage during the measurements.

Rotational measurements with controlled shear rate (CSR) were performed. The measurements were made in the shear rate range of 0.5–300 s^{-1} followed by a downwards curve in order to evaluate the existence of thixotropy. It should be highlighted that the shear–strain rate range tested corresponded to the typical shear rate range during the grout injection process in order to reveal representative rheological parameters. This shear–strain rate range was associated with the pressures usually adopted during the consolidation of old masonries, which must be below 0.5–1 bar in order to not cause instability of the masonry [28].

A common approach to determine the yield stress is fitting a yield-stress-containing model to shear stress vs. shear–strain rate data. Over the years, useful progress has been made in applied rheology and different rheological models have been suggested [32–38]. The most popular rheological models that have been proposed to describe suspensions with yield stress are the Bingham, Casson and Herschel–Bulkley [3,35–37]. However, there is no single model that can adequately describe the behavior of all complex suspensions, in addition to the fact that none of these models have a limit on the maximum shear stress. Most models assume that the shear stress increases infinitely with the shear–strain rate, which does not reflect reality, since any fluid has a maximum shear stress value. To overcome this limitation, Vipulanandan and Mohammed [13] proposed a hyperbolic model that has been successfully used in the characterization of drilling muds [33,34]. However, for the materials used in this study and for the range of shear–strain rates experienced (i.e., below the maximum shear stress of the grout due to structural safety requirements of masonry) the Herschel–Bulkley proved to be the model that best described the behavior of NHL-based grouts, corroborating what has been concluded in previous studies [36,37]. The Herschel–Bulkley model is able to predict the yield stress at low shear rates and determine the shear-thinning behavior of cementitious suspensions [37]. Thus, the experimental data (shear stress vs. shear–strain rate) were adjusted to the Herschel–Bulkley equation (see Equation (1)) to calculate the yield stress.

$$\tau = \tau_0 + k\dot{Y}^n \tag{1}$$

where τ_0 is yield stress (Pa), k is the correlation parameter (Pa.s^n), $\dot{Y}$ is the shear–strain rate (s^{-1}) and n is the flow index (-), which describes shear thinning ($n < 1$) and shear thickening behavior ($n > 1$).

Moreover, controlled shear stress (CSS) was adopted to carry out the stress ramp tests. The stress was increased from 0.006 to 140 Pa and followed by a down ramp with a linear ramp of 0.3 Pa/s. The subsequent apparent viscosity and shear rate were measured for 50 min. It is known that, when a grout exceeds the yield stress, an abrupt and profound change in their microstructure is observed, leading to a state of less resistance. This microstructure change can be graphically portrayed in plots of apparent viscosity against shear stress, where it is possible to verify that below a critical shear stress the fluid in question appeared to have an infinite viscosity, and above the critical stress a shear-thinning behavior. All the rheological measurements were carried out with a constant temperature of 20 °C, maintained by means of a temperature unit control. A solvent trap was used to prevent drying of the grout samples during testing.

3. Results and Discussion

3.1. Thixotropy and Yield Stress by CSR Method

In general, a yield stress fluid does not depend on shear history and a linear proportionality between the shear stress and the shear–strain rate can be established. However, although the NHL-based grout exhibits yield stress, its behavior is time-dependent and a non-linear relationship between the shear stress and shear–strain rate is often observed [2,11]. Considering these statements, a series of controlled shear–strain rate measurements was carried out on the same sample for 50 min with 10 min intervals between each measurement. This is shown in Figure 2 for the case of the grout without PP fibers.

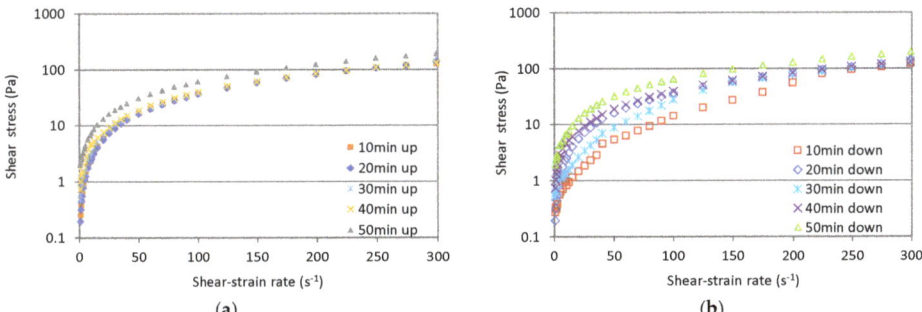

Figure 2. Shear stress curves of the reference grout: (**a**) up curve, (**b**) down curve.

From Figure 2, it can be seen that the shear stress constantly increases with time, which is a consequence of the hydration process of NHL. Additionally, the characteristic behavior of a thixotropy material, in which the up curve has higher shear stresses values than the down curve, can also be noted. Nevertheless, the thixotropy tends to become less pronounced over time, which can be associated with the formation of hydration products that are not destroyed by the application of the shear–strain rate [38,39]. In Figures 3 and 4, the results for the grouts with PP fibers are represented. As expected, the PP fibers had an influence on the rheological curves of fresh grouts by increasing their shear-thinning behavior [40]. It can be noted that the addition of the PP fibers increased the measured shear stress of grouts, especially for the dosage of 0.1%, at low shear–strain rates (see Figure 4).

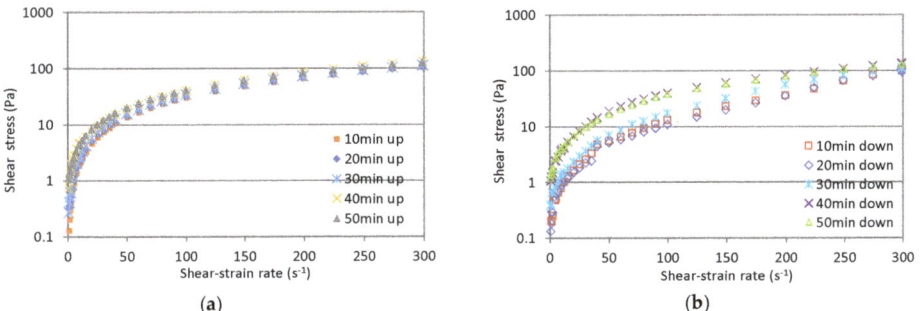

Figure 3. Shear stress curves of grout with 0.03% polypropylene (PP) fibers: (**a**) up curve, (**b**) down curve.

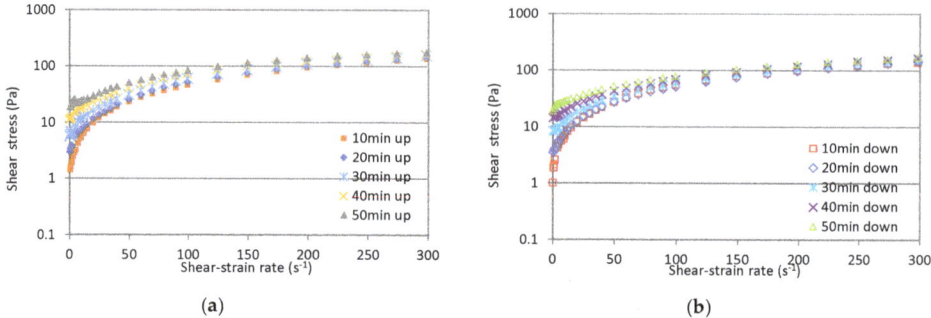

Figure 4. Shear stress curves of grout with 0.1% PP fibers: (**a**) up curve, (**b**) down curve.

The increase in shear stress and the consequent loss of workability resulting from the increase in fiber content was also obtained in studies by Banfill et al. [41] and Tabata-Baeian et al. [42]. This occurs due to mechanical interlocks and bridges between binder particles and fibers. The results show that the effect of thixotropy is less pronounced in the presence of PP fibers and becomes almost non-existent for grouts with a higher fiber dosage, which reflects a nondependent shearing history caused by a high volume of PP fibers. However, as time passes, the thixotropy becomes less evident, as happened with the reference grout. Based on the Herschel–Bulkley model, two yield stresses were calculated, namely the static and dynamic yield stress for the up and down curves, respectively (see Figure 5). Since the samples were resting prior to measurement, the up curve is supposed to provide the static yield stress while the down curve provides the dynamic yield stress [11].

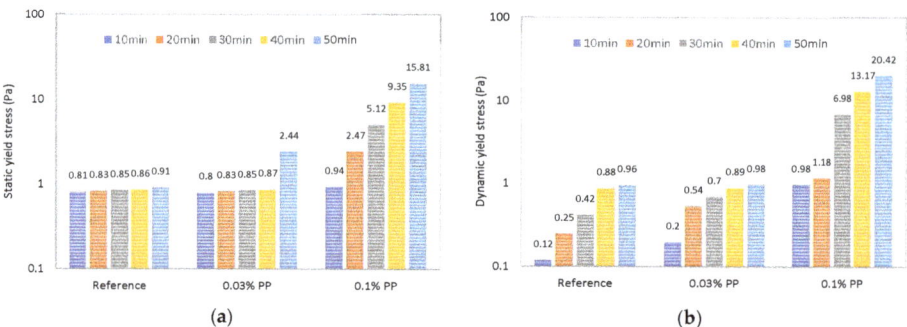

Figure 5. Effects of PP fibers on yield stress values by controlled shear rate (CSR) measurement: (**a**) static yield stress, (**b**) dynamic yield stress.

From Figure 5, it can be observed that the yield stress gradually increases with the addition of PP fibers, which is consistent with References [40,41]. However, there is a greater increase in yield stress values between the dosage of 0.03% and 0.1%. This occurred because the interconnections between the suspension systems were modified by adding PP fibers [37,40]. This means that the fibers fill the gaps between binder particles, which contributes to the formation of a more cohesive microstructure that results in higher yield stress [43]. Notwithstanding, it can be seen from Figure 5b that the dynamic yield stress is lower than the static one for very low contents of PP fibres. On the other hand, for the highest fiber content, it is possible to observe that for time periods above 30 min, the dynamic yield stress is higher than the static one, which may be due to the combined effects of the grout hydration and the very cohesive microstructure.

3.2. Yield Stress by CSS Method

CSS measurements were performed due to their wide-ranging suitability for the determination of yield stress [9,10]. So, as previously described, the grout samples were subjected to a CSS from 0.006 to 140 Pa for the up curve and down curve. As shown in Figure 6, for the grout with 0.03% PP fibers, the change in the apparent viscosity in the up curve before yielding and after yielding is rather sudden; i.e., the apparent viscosity has a high value at low shear stresses and tends to a low constant viscosity value at higher shear stresses.

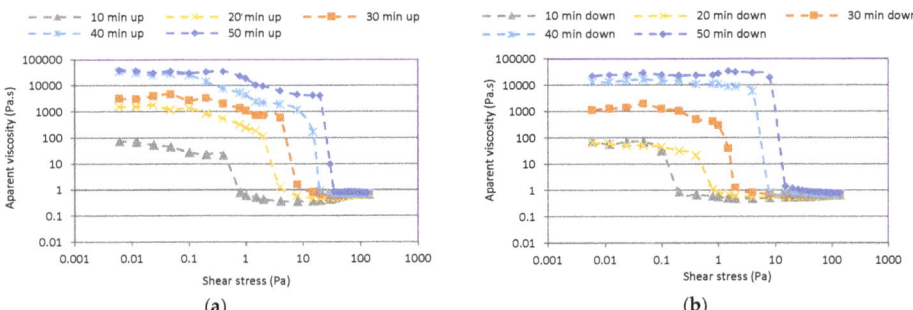

Figure 6. Curves obtained from the shear stress ramp for grout with 0.03% PP fibers: (**a**) up curves (**b**) down curves.

Based on the results presented in Figure 6, it can be noticed that the up curve shows a higher apparent viscosity and yield stress than the down curve due to the thixotropy. The other grout compositions showed an analogous trend. The yield stresses increased with time from 0.4 to 20 Pa. The yield stress increased to 4 Pa after 20 min, and between 20 and 40 min the yield value increased from 4 to 15 Pa. A comparison of the static and dynamic yield stress results for all the grout compositions is shown in Figure 7. As predicted, the static yield stress values are higher compared to the dynamic yield stress. This can be explained by the particle bonds that are broken down due to the shearing from the up ramp, which reduces the dynamic yield stress values. An interesting fact is that, even though both yield stresses increase with the addition of fibers, the yield stresses appear to increase more significantly over time. This can be justified by the hydration reactions of NHL, since during the hydration process a build-up of the microstructure occurs and, consequently, the yield stress increases [38,39].

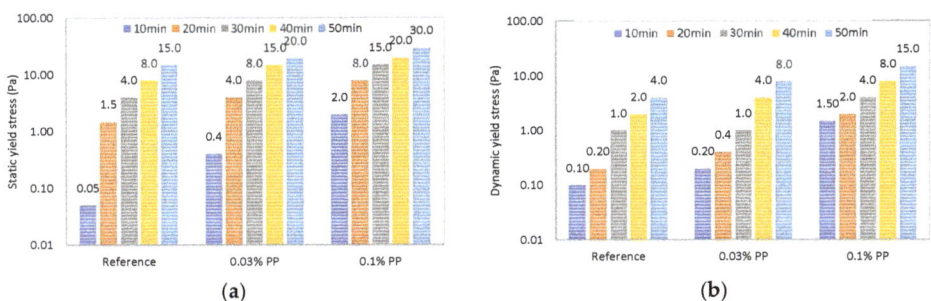

Figure 7. Effects of PP fibers on yield stress values by controlled shear stress (CSS) measurement: (**a**) static yield stress, (**b**) dynamic yield stress.

3.3. Comparison between Yield Stress, Critical Shear–Strain Rate and Time Period

In Figure 8, the yield stress values obtained by the fitting of the Herschel–Bulkley model to the experimental data obtained by CSS and CSR methods are compared. From these results, it can be noted that the yield stress values show just moderate differences between both measurement methods; however, the CSR leads to smaller values than the CSS. There are several reasons for this difference, such as the internal cohesion force of the sample's microstructure and the sample shear history. However, it is known that measurements taken with CSS often provide better results than measurements with controlled CSR [44]. Nevertheless, the results confirm the efficiency of the Herschel–Bulkley model to determine the yield stress of the NHL grouts, regardless of the measurement method used [45].

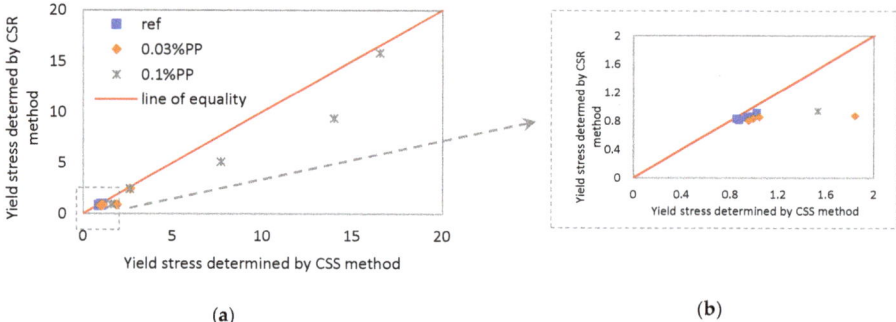

Figure 8. Comparison of static yield stress modeled by CSS and CSR: (**a**) all yield stress range; (**b**) focus on lower yield stress values.

In order to highlight the ability of the different measurement methods to characterize NHL grouts, a comparison between the experimental curves of the CSS and CSR methods at 20 min for the reference grout and the one with 0.03% PP fibers are shown in Figures 9 and 10, respectively (the other time periods and compositions presented similar behavior). Based on the results obtained, it can be seen that regardless of the measurement method performed, the flow curves show similar behavior. However, the addition of PP fibers caused a slight difference between the CSS and CSR curves at high shear–strain rates (see Figure 10).

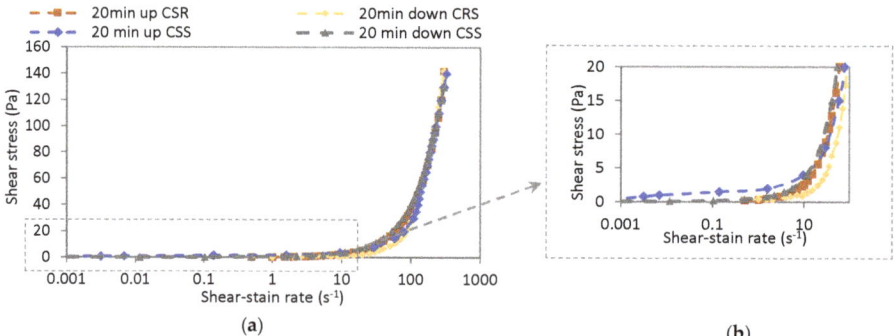

Figure 9. Flow curve obtained by CSS and CRS methods for the reference grout: (**a**) all shear-strain rate range; (**b**) focus on lower shear-strain rates.

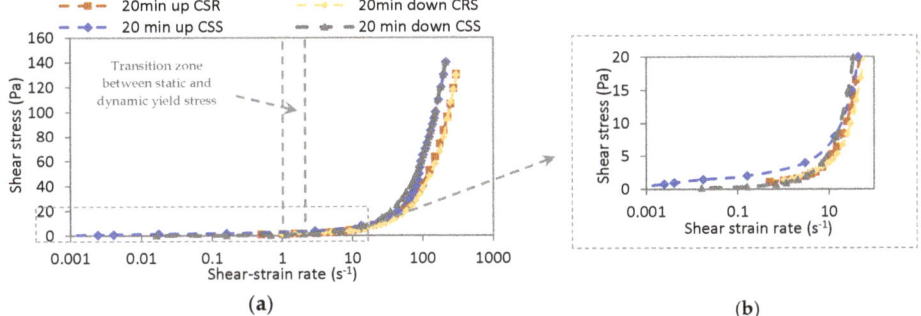

Figure 10. Flow curve obtained by CSS and CRS methods for grout with 0.03% PP fibers: (**a**) all shear-strain rate range; (**b**) focus on lower shear-strain rates.

From the results presented in Figure 10, a slight linear increase in shear stress can be noted until 1 s^{-1} for both the up and down curves. Moreover, a considerable increase in shear stress can be observed for shear–strain rates above 3 s^{-1}, which explains the change of apparent viscosity presented in Figure 6. Based on these results, it can be seen that the shear–strain rate range of between 1 and 3 s^{-1} is the transition zone between the two yield stresses, or, in other words, a shear–strain rate of up to 1 s^{-1} would lead to a static yield stress, while shear–strain rates higher than 3 s^{-1} would provide a dynamic yield stress. It should be noted, however, that this transition zone between the two different yield stresses is time-dependent, as shown in Table 3.

Table 3. Evolution of shear–strain rate range for the yield stress transition as a function of time.

Measurement Instant	Shear–Strain Rate Range of Transition Zone [1]
10 min	0.1–1.0 s^{-1}
20 min	1.0–3.0 s^{-1}
30 min	5.0–9.0 s^{-1}
40 min	8.0–12.0 s^{-1}
50 min	11.0–17.0 s^{-1}

[1] shear–strain rates presented in each instance encompass all grout compositions studied.

By analyzing the results of Table 3, it is possible to conclude that the shear–strain rate, which is necessary for the grout to start to flow or to continue flowing (depending on the case), increases significantly over time. Increases in the critical shear–strain rate of up to 60% and more than 90% between 10–20 min and 10–30 min were found, respectively. This behavior can be seen as a consequence of hydration reactions [46], which can significantly affect the success of the masonry consolidation operation if proper precautions are not taken during the grouting design, such as adjusting the grout's yield stress to the shear–strain rate range to which the grout will be subjected during injection.

From a practical point of view, the grout is in a broken-down state when it reaches the masonry core in the first moments of the injection operation; however, when the grout is at a distance significantly away from the injection point, the maximum yield stress may be reached due to the reduction of the shear–strain rates. So, as previously highlighted, the yield stress value must be chosen according to the shear–strain rate range of interest. In this sense, a dynamic yield stress should be used at the beginning of the injection operation when the grout is in a fully broken-down state or, in other words, subjected to high shear rates. Meanwhile, static yield stress should be a design parameter only for lower shear–strain rates (i.e., at later stages of the grouting operation) when the links between NHL particles start to take place. This is also effective in situations involving sudden stoppages of the injection process. In order to better illustrate the relationship between time and the shear–strain rate range that causes the transition between static and dynamic yield stress, a regression analysis was made. Shear–strain rate values for all grout compositions were achieved by using the CSS method and measurement instances of 10, 20, 30, 40 and 50 min were considered. As shown in Figure 11, a good correlation was obtained ($r^2 > 0.9$) and two equations were proposed in order to allow the estimation of critical shear–strain rate values based on the time period after the grout mixing process was completed.

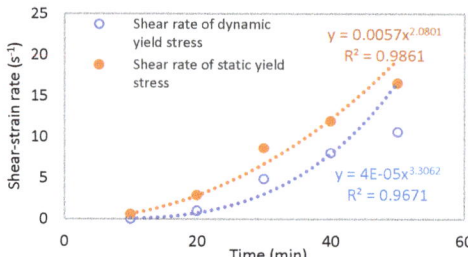

Figure 11. Correlation between the shear-strain rate range and time.

In addition, correlations between yield stresses and shear–strain rates were also established (see Figure 12). Since good correlations between the yield stress values and shear–strain rate were found, several equations were proposed. In this way, the shear–strain rate values could be forecasted by knowing the yield stress values and, therefore, predications of the critical shear–strain rate below which there is a transition from dynamic to static yield stress could be made. It should be highlighted that the validity domain of the proposed models is limited to a given set of materials and assumptions, so any extrapolations must be done carefully.

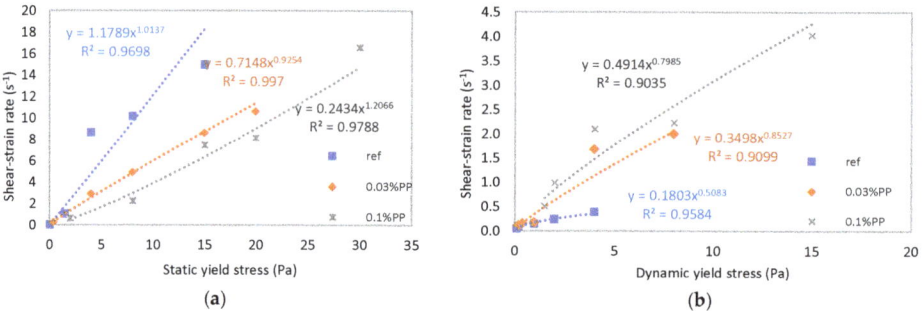

Figure 12. Correlation between the shear–strain rate range and (**a**) static yield stress, (**b**) dynamic yield stress.

The results and measurements performed show that these methods enable the evaluation of the yield stress of NHL-based grouts and the critical shear–strain rate range. In addition, to maximize the injectability of the grout, it is imperative not to stop the grouting operation for a long time (for instance time periods greater than 20 min) in order to prevent the yield stress from increasing. To restart the flow, it is necessary to increase the shear–strain rate or, in other words, increase the injection pressure, which can consequently cause additional damage to the masonry.

4. Conclusions

In this study, the yield stress of NHL-based grouts over a range of time periods and with different polypropylene fiber contents was investigated. The grouts were made with a polycarboxylate powder superplasticizer and PP fibers in amounts of up to 0.1% of PP fiber content (by weight). Yield stress is a key parameter in the design and optimization of grouts for the consolidation of old masonries. Two yield stress values, namely the static and dynamic yield stress, were determined based on different experimental measurement methods. The results obtained allow us to draw the following conclusions:

1. The static yield stress values were higher compared to the dynamic yield stress. Differences between both yield stresses were obtained in a range of 33% to 670%, depending on the content of PP fibers. This can be explained by the semi-disturbed state of the grout's microstructure when the dynamic yield stress was determined, which was mainly due to the shear history dependence of the grout.
2. The PP fibers influenced on the rheology of the NHL grouts by increasing their shear-thinning behavior. Moreover, both yield stress values increased with the presence of PP fibers. For example, for the reference grout (without fibers), the minimum and maximum values of the static yield stress varied from 0.05 to 15 Pa, while for the grout with 0.1% PP fibers, the values varied between 2 and 30 Pa. The amount of changes in the yield stress values was due to the structural build up and flocculation, which was a consequence of the mechanical interlocks between the NHL particles and fibers.
3. The yield stress values depended on the measuring method. However, the yield values determined by CSS and CSR only showed moderate differences between them, which confirmed the efficiency of the Herschel–Bulkley model in determining the yield stress of NHL grouts whatever the measurement method used.
4. The results and measurements performed showed that these methods enable us to evaluate the existence of a critical shear–strain rate range of 0.1 to 17.0 s^{-1} (depending on time period). Below this, there was a transition between dynamic and static yield stress.
5. The dynamic yield stress should be used as design parameter at early stages of the grout injection process, whilst at a later stage, when the shear rate is slowing down, the static yield stress should be considered.
6. Several equations that allow the estimation of the critical shear–strain rate as a function of time and yield stress have been proposed in order to promote better design of injection grouts.
7. The critical shear–strain rate range increases over time. Therefore, to maximize the injectability of the grout, it is imperative to avoid stopping the grouting operation for periods longer than 20 min.

The findings from this study are relatively promising for further understanding the evolution of yield stress of NHL grouts as a function of time, shear–strain rate and PP fiber content. Furthermore, the results presented contribute helpful information on the design input of grouting operations.

Author Contributions: Conceptualization, L.G.B., F.M.A.H. and M.T.C.; methodology, L.G.B.and M.T.C.; writing—original draft preparation, L.G.B.; writing—review and editing, L.G.B., M.T.C. All authors have read and agreed to the published version of the manuscript.

Funding: This work is funded by National Funds through FCT/MCTeS—Portuguese Foundation for Science and Technology, Reference UID/CTM/50025/2013 and FeDeR funds through the COMPeTe 2020 Programme under the project number POCI-01-0145-FeDeR-007688.

Conflicts of Interest: The authors declare no conflict of interest.

References

1. Kalagri, A.; Miltiadou-Fezans, A.; Vintzileou, E. Design and evaluation of hydraulic lime grouts for the strengthening of stone masonry historic structures. *Mater. Struct.* **2010**, *43*, 1135–1146. [CrossRef]
2. Baltazar, L.G.; Henriques, F.M.A.; Cidade, M.T. Rheological characterization of injection grouts using rotational rheometry. In *Advances in Rheology Research*; Nova Science Publishers: New York, NY, USA, 2017; pp. 13–42, ISBN 978-1-53612-876-5.
3. Baltazar, L.G.; Henriques, F.M.A.; Cidade, M.T. Rheology of natural hydraulic lime grouts for conservation of stone masonry—Influence of compositional and processing parameters. *Fluids* **2019**, *4*, 13. [CrossRef]
4. Barnes, H.A.; Walters, K. The yield stress myth? *Rheol. Acta* **1985**, *24*, 323–326. [CrossRef]

5. Hartnett, J.P.; Hu, R.Y.Z. Technical note: The yield stress—An engineering reality. *J. Rheol.* **1989**, *33*, 671. [CrossRef]
6. Schurz, J. The yield stress—An empirical reality. *Rheol. Acta* **1990**, *29*, 170–171. [CrossRef]
7. Keentok, M. The measurement of the yield stress of liquids. *Rheol. Acta* **1982**, *21*, 325–332. [CrossRef]
8. Dzuy, N.Q.; Boger, D.V. Yield stress measurement for concentrated suspensions. *J. Rheol.* **1983**, *27*, 321. [CrossRef]
9. Coussot, P.; Nguyen, Q.D.; Huynh, H.T.; Bonn, D. Viscosity bifurcation in thixotropic, yielding fluids. *J. Rheol.* **2002**, 573–589. [CrossRef]
10. Barnes, H.A. The yield stress—A review or 'παντα ρει'—Everything flows? *J. Non Newton Fluid* **1999**, *81*, 133–178. [CrossRef]
11. Hakansson, U. Rheology of Fresh Cement-Based Grouts. Ph.D. Thesis, The Royal Institute of Technology, Stockholm, Sweden, 1993.
12. Al-Martini, S.; Nehdi, M. Effect of chemical admixtures on rheology of cement pastes at high temperature. *J. ASTM Int.* **2007**, *4*, 1–17. [CrossRef]
13. Vipulanandan, C.; Mohammed, A.S. Hyperbolic rheological model with shear stress limit for acrylamide polymer modified bentonite drilling muds. *J. Petrol Sci. Eng.* **2014**, *122*, 38–47. [CrossRef]
14. Collepardi, M. Degradation and restoration of masonry walls of historical buildings. *Mater. Struct.* **1990**, *23*, 81–102. [CrossRef]
15. Pozo-Antonio, J.S. Evolution of mechanical properties and drying shrinkage in lime-based and lime cement-based mortars with pure limestone aggregate. *Constr. Build. Mater.* **2015**, *77*, 472–478. [CrossRef]
16. Puertas, F.; Amat, T.; Fernández-Jiménez, A.; Vázquez, T. Mechanical and durable behaviour of alkaline cement mortars reinforced with polypropylene fibres. *Cem. Concr. Res.* **2003**, *33*, 2031–2036. [CrossRef]
17. Banthia, N.; Gupta, R. Influence of polypropylene fiber geometr on plastic shrinkage cracking in concrete. *Cem. Concr. Res.* **2006**, *36*, 1263–1267. [CrossRef]
18. Noushini, A.; Samali, B.; Vessalas, K. Effect of polyvinyl alcohol (PVA) fibre on dynamic and material properties of fibre reinforced concrete. *Constr. Build. Mater.* **2013**, *49*, 374–383. [CrossRef]
19. Liu, J.; Sun, W.; Miao, C.; Liu, J.; Li, C. Assessment of fiber distribution in steel fiber mortar using image analysis. *J. Wuhan Univ. Technol. Mater. Sci.* **2012**, *27*, 166–171. [CrossRef]
20. Campello, E.; Pereira, M.V.; Darwish, F. The effect of short metallic and polymeric fiber on the fracture behavior of cement mortar. *Procedia Mater. Sci.* **2014**, *3*, 1914–1921. [CrossRef]
21. Lucolano, F.; Liguori, B.; Colella, C. Fibre-reinforced lime-based mortars: A possible resource for ancient masonry restoration. *Constr. Build. Mater.* **2013**, *38*, 785–789. [CrossRef]
22. Di Bella, G.; Fiore, V.; Galtieri, G.; Borsellino, C.; Valenza, A. Effect of natural fibers reinforcement in lime plasters (kenaf and sisal vs. polypropylene). *Constr. Build. Mater.* **2014**, *58*, 159–165. [CrossRef]
23. Chan, R.; Bindiganavile, V. Toughness of fibre reinforced hydraulic lime mortar. Part 1: Quasi-static response. *Mater. Struct.* **2010**, *43*, 1435–1444. [CrossRef]
24. Chan, R.; Bindiganavile, V. Toughness of fibre reinforced hydraulic lime mortar. Part 2: Dynamic response. *Mater. Struct.* **2010**, *43*, 1445–1455. [CrossRef]
25. Barbero-Barrera, M.M.; Medin, N.F. The effect of polypropylene fibers on graphite-natural hydraulic lime pastes. *Constr. Build. Mater.* **2018**, *184*, 591–601. [CrossRef]
26. CEN. *Building Lime-Part 1: Definitions, Specifications and Conformity Criteria*; EN 459-1; CEN, European Committee for Standardization: Brussels, Belgium, 2010.
27. Binda, L.; Saisi, A.; Tedeschi, C. Compatibility of materials used for repair of masonry buildings: Research and applications. *Fracture Failure Nat. Build. Stone* **2006**, 167–182. [CrossRef]
28. Jorne, F.; Henriques, F.M.A.; Baltazar, L.G. Influence of superplasticizer, temperature, resting time and injection pressure on hydraulic lime grout injectability. Correlation analysis between fresh grout parameters and grout injectability. *J. Build. Eng.* **2015**, *4*, 140–151. [CrossRef]
29. Baltazar, L.G.; Henriques, F.M.A.; Cidade, M.T. Experimental study and modeling of rheological and mechanical properties of NHL grouts. *J. Mater. Civil Eng.* **2015**, *27*. [CrossRef]
30. ASTM. *Standard Specification for Chemical Admixtures for Concrete*; ASTM C494; ASTM International: West Conshohocken, PA, USA, 2005.

31. Baltazar, L.G.; Henriques, F.M.A.; Cidade, M.T. Combined effect of silica fume and nanosilica on the performance of injection grouts for consolidation of heritage buildings. In *Advances in Rheology Research*; Nova Science Publishers: New York, NY, USA, 2019; Volume 32, pp. 197–238, ISBN 978-1-53616-684-2.
32. Yoshimura, A.S. A comparison of techniques for measuring yield stresses. *J. Rheol.* **1987**, *31*, 699. [CrossRef]
33. Sugiura, J.; Samuel, R.; Oppelt, J.; Ostermeyer, G.P.; Hedengren, J.; Pastusek, P. Drilling modeling and simulation: Current state and future goals. *J. Pet. Technol.* **2015**, *67*, 140–142. [CrossRef]
34. Mohammed, A.; Mahmood, W.; Ghafor, K. TGA, rheological properties with maximum shear stress and compressive strength of cement-based grout modified with polycarboxylate polymers. *Constr. Build. Mater.* **2020**, *235*, 117534. [CrossRef]
35. Bala, M.; Zentar, R.; Boustingorry, P. Comparative study of the yield stress determination of cement pastes by different methods. *Mater Struct.* **2019**, *52*, 102. [CrossRef]
36. Vance, K.; Sant, G.; Neithalath, N. The rheology of cementitious suspensions: A closer look at experimental parameters and property determination using common rheological models. *Cem. Concr. Compo.* **2015**, *59*, 38–48. [CrossRef]
37. Jiao, D.; Shi, C.; Yuan, Q.; Zhu, D.; De Schutter, G. Effects of rotational shearing on rheological behavior of fresh mortar with short glass fiber. *Constr. Build. Mater.* **2019**, *203*, 314–321. [CrossRef]
38. Jiao, D.; Shi, C.; Yuan, Q. Time-dependent rheological behavior of cementitious paste under continuous shear mixing. *Constr. Build. Mater.* **2019**, *226*, 591–600. [CrossRef]
39. Mostafa, A.M. Physical and Chemical Kinetics of Structural Build-Up of Cement Suspensions. Ph.D. Thesis, Université de Sherbrooke, Sherbrooke, QC, Canada, 2016.
40. Zhang, K.; Pan, L.; Li, J.; Lin, C.; Cao, Y.; Xu, N.; Pang, S. How does adsorption behavior of polycarboxylate superplasticizer effect rheology and flowability of cement paste with polypropylene fiber? *Cem. Concr. Compos.* **2019**, *95*, 228–236. [CrossRef]
41. Banfill, P.F.G.; Starrs, G.; Derruau, G.; McCarter, W.J.; Chrisp, T.M. Rheology of low carbon fibre contente reinforced cement mortar. *Cem. Concr. Compos.* **2006**, *28*, 773–780. [CrossRef]
42. Tabatabaeian, M.; Khaloo, A.; Joshaghani, A.; Hajibandeh, E. Experimental investigation on effects of hybrid fibers on rheological, mechanical, and durability properties of high-strength SCC. *Constr. Build. Mater.* **2017**, *147*, 497–509. [CrossRef]
43. Beigi, M.H.; Berenjian, J.; Omran, O.L.; Nik, A.S.; Nikbin, I.M. An experimental survey on combined effects of fibers and nanosilica on the mechanical, rheological, and durability properties of self-compacting concrete. *Mater. Des.* **2013**, *50*, 1019–1029. [CrossRef]
44. Mezger, T.G. *The Rheology Handbook: For Users of Rotational and Oscillatory Rheometers*; Vincentz Network: Hannover, Germany, 2006; ISBN 978-3-87870-174-3.
45. De Larrard, F.; Ferraris, C.; Sedran, T. Fresh concrete: A Herschel-Bulkley material. *Mater. Struct.* **1998**, *31*, 494–498. [CrossRef]
46. Benyounes, K. Rheological behavior of cement-based grout with Algerian bentonite. *SN App. Sci.* **2019**, *1*, 1037. [CrossRef]

© 2020 by the authors. Licensee MDPI, Basel, Switzerland. This article is an open access article distributed under the terms and conditions of the Creative Commons Attribution (CC BY) license (http://creativecommons.org/licenses/by/4.0/).

Article

Rheology Methods as a Tool to Study the Impact of Whey Powder on the Dough and Breadmaking Performance of Wheat Flour

Christine Macedo, Maria Cristiana Nunes *, Isabel Sousa and Anabela Raymundo

LEAF-Linking Landscape, Environment, Agriculture and Food, Instituto Superior de Agronomia, Universidade de Lisboa, Tapada da Ajuda, 1349-017 Lisboa, Portugal; chmacedo1309@gmail.com (C.M.); isabelsousa@isa.ulisboa.pt (I.S.); anabraymundo@isa.ulisboa.pt (A.R.)
* Correspondence: crnunes@gmail.com

Received: 28 February 2020; Accepted: 11 April 2020; Published: 14 April 2020

Abstract: Considering the nutritional value, whey is an excellent ingredient for the development of food products, in line with the concept of a circular economy for the reuse of industry by-products. The main objective of this work was to evaluate the impact of the whey addition on the rheology of wheat flour dough and breadmaking performance, using both empirical and fundamental methods. Different levels of commercial whey powder (0%, 12%, 16% and 20% w/w) were tested in a bread formulation previously optimized. Dough mixing tests were performed using Micro-doughLab and Consistograph equipment, to determine the water absorptions of different formulations and evaluate empirical rheology parameters related to mixing tolerances. Biaxial extension was applied by the Alveograph to simulate fermentation during the baking process. Fermented doughs were characterized in a Texturometer using penetration and extensibility tests, and by small amplitude oscillatory shear (SAOS) measurements, a fundamental rheology method, in a Rheometer applying frequency sweeps. Loaf volume and firmness were used to study the breadmaking quality. Despite a negative impact on the empirical rheology parameters of the dough and poorer baking results, the use of this by-product should be considered for nutritional and sustainability reasons. In addition, significant correlations ($r^2 > 0.60$) between the dough rheology parameters obtained from the empirical measurements were established. Changes in the gluten structure were not accurately detected by the SAOS measurements and Texture Profile Analysis of the doughs, and a correlation between fundamental and empirical measurements was not found. Consistograph or Micro-doughLab devices can be used to estimate bread firmness. Extensional tests in the Texturometer, using SMS/Kieffer Dough and Gluten Extensibility Rig, may predict loaf volume.

Keywords: bread; whey; complex fluids; experimental rheology; breadmaking

1. Introduction

Considering the large amounts of whey produced all the years by the cheese industry, coupled with its high organic matter composition, namely lactose and proteins, leading to high chemical oxygen demand when disposed into the effluents, whey has been considered an important pollution problem and several strategies have been developed to add value to this by-product, including bringing it back to the food value-chain, as in circular economy principles [1,2].

As whey contains some important components, such as lactose, proteins and minerals, it is recognized as a valuable source of high-grade proteins, mainly β-lactoglobulin and α-lactalbumin, which constitute ca. 50% and 20% of the total protein content, respectively; the remainder is accounted for by immunoglobulins, bovine serum albumin, protease peptones and other minor proteins [3,4]. The excellent functional properties of whey proteins have been recognized, namely gelation and binding properties; therefore, whey is widely used as a functional ingredient in many formulated bakery and dairy foods [2,5,6]. In addition to proteins, cheese whey is also rich in lactose; thus, its biotechnological value as a fermentation substrate has also been explored, namely to produce bioethanol, biogas and lactic acid [2,7,8].

There are several studies about the incorporation of whey in traditional bread [9–17], revealing an increase of total mineral content, calcium, magnesium, phosphorus, potassium and zinc contents and also lactose and lactic acid, followed by a positive effect on crust color, sweet and yeast flavor, expressed as a positive sensory impact. However, a negative impact on the development of the gluten matrix was also verified, expressed as a softening in the dough system, by the reduction of the viscoelastic moduli values [16,17]. More recently, the use of whey proteins on the mimetic effect of gluten, for the development of gluten-free dough, has also been investigated [18–22].

The incorporation of other protein sources into wheat flour is a market tendency, but it is a major technological challenge since the wheat flour gluten-forming proteins (glutenin and gliadin), responsible for the viscoelastic dough structure, may be perturbed either by a dilution effect of the gluten proteins or/and by interfering in the intermolecular linkages of the protein matrix. This matrix is the three-dimensional network formed by gluten, which surrounds the starch granules and retains the air incorporated during the mixing process and the carbon dioxide (CO_2) produced by yeast fermentation. If a large amount of lactose, proteins, or fibers is incorporated in wheat flour, the amount of water necessary to obtain the desirable mechanical properties of the dough changes [16,23,24]. The mineral content of whey also influences the dough formation properties of gluten proteins and may improve both association or dissociation of dough components [10].

Many rheology devices have been used by cereal technologists and researchers to predict the flour performance throughout the whole bread processing—during mixing and kneading, fermentation, molding, fermentation and baking steps. For this, empirical techniques are still indispensable and recognized as standard methods [25]. The empirical methods include the following instruments: Farinograph, Mixograph, Extensography, Alveography, Amylograph, Mixolab and Texturogram. Fundamental rheology techniques are also used to study flour dough systems, by means of static or dynamic measurements. The dynamic oscillatory rheology involving small deformations (SAOS) is being preferred to study the structural and fundamental properties of the wheat dough [26,27]. Both fundamental and empirical methods present some disadvantages. In the baking area, the empirical rheology methods are especially prominent to study the influence of flour constituents, and additives, on dough behavior, and are commonly used by the industry.

This work is part of a project that aims to optimize the healthy bread composition and technological processing by determining the maximum content of whey to be incorporated in traditional bread production, keeping the mechanical behavior of the dough and sensory appealing of the resulting bread. Four types of empirical instruments, Micro-doughLab, Consistograph, Alveograph and Texturometer, were employed to study the rheology properties of wheat dough enriched with whey powder. Fundamental SAOS measurements were evaluated as well, and the obtained results from these different types of tests are discussed and compared to estimate dough performance during processing and future bread properties. Bread quality properties like texture (firmness) and volume were also determined.

2. Material and Methodology

2.1. Materials

The ingredients used for the preparation of the bread doughs were the following: wheat flour (Granel T65, Portugal) with a minimum of 8.0% gluten (db) and falling number higher than 220 s, commercial whey powder (Lactogal S.A., Portugal), dehydrated yeast (Fermipan, France), commercial sugar and salt, SSL-E481-sodium stearoyl-2 lactylate (Puratos, Portugal) and distilled water. Whey powder has 74.0 g of lactose, 12.0 g protein, 1.4 g lipids and 1.3 g of salt per 100 g of the product, with a moisture content of 11.3 g/100 g, considering the information provided by the supplier.

2.2. Methodology

Wheat flour was partially replaced by different amounts of whey powder—0% (Control), 12%, 16% and 20% w/w. The amount of water was added according to the values found in Micro-doughLab mixing tests, at a 14% moisture basis, as shown in Table 1. The moisture content of the flour and whey powder was determined in a moisture infra-red determination balance (ADAM PMB 202).

Table 1. Bread dough samples formulation and respective codes.

Ingredients (g/100 g)	Doughs			
	Control (C)	12% Whey (12D)	16% Whey (16D)	20% Whey (20D)
Wheat flour	100.0	88.0	84.0	80.0
Whey powder	0.0	12.0	16.0	20.0
Water Absorption (14% moisture basis)	52.2	39.4	38.2	36.4

For texture and oscillatory measurements, doughs were prepared using a bread formulation. The other ingredients were the same for all formulations: yeast (4.0 g), salt (1.7 g), sugar (1.0 g) and SSL (0.5 g) in relation to 100 g of wheat flour + whey powder, according to a previously optimized formulation [23,24].

The preparation of the bread doughs was carried out in a thermal processor (Bimby-Vorwerk, Carnaxide, Portugal). First, yeast was activated in warm water in the processor cup, for 30 s at position 3 at 37 °C. The solid ingredients were added and homogenized during 60 s in position 6, and subsequent kneading during 120 s. The dough was placed in a rectangular bread container (5.0 cm × 20.0 cm × 8.5 cm), previously sour and floured, followed by fermentation during 60 min at 37 °C (optimum time/temperature of yeast activity previously optimized) in an electric oven (Arianna XLT133, Cadoneghe, Italy). After fermentation, doughs were characterized by means of texture assays and fundamental oscillatory rheology. For breadmaking tests (loaf volume and crumb firmness), the dough was baked at 160 °C for 30 min. Breads were analyzed after cooling for 2 h. Three loaves of each formulation were prepared.

2.2.1. Micro-doughLab

The Micro-doughLab 2800 (Perten Instruments, Sidney, Australia) was used to determine the optimum water absorption capacity to reach a peak of 130 mN.m, using 4.00 ± 0.01 g sample at 14% moisture basis, according to the AACC method 54.70-01: High-speed Mixing Rheology of Wheat Flour Using the DougLab, modified for Micro-doughLab. Sample and water weights were corrected from sample moisture content. Manufacturer's rapid mixing protocol was used, mixing the wheat flour and whey at a constant 120 rpm speed and temperature of 30 °C for 10 min. Measurements were repeated at least three times for each sample. As a result, a mixing curve was obtained that provides the dough's mixing properties—peak resistance (mN.m), dough development time (s), stability (s), softening (mN.m) and peak energy (Wh/kg).

2.2.2. Consistograph

The Consistograph (AlveoLAB, Chopin Technologies, Cedex, France) was used to determine the water absorption capacity and the physical properties of the wheat flour/whey dough systems, according to the AACC 54–50.01 method. First, an amount of water, based on the initial moisture content of the flour + whey, is added in order to reach a constant hydration level (76.47% moisture on a dry-matter basis). The peak pressure recorded during kneading is used to calculate the water absorption of the flour sample at a target pressure of 2200 ± 100 mm H_2O. Then, the subsequent test is performed at the adapted hydration level previously determined. Maximum pressure (mbar), time to reach maximum pressure (s), tolerance to kneading (s), and consistency of the dough after 250 and 450 s (mbar) were obtained. Measurements were repeated at least three times for each sample.

2.2.3. Alveograph

To determine the resistance of doughs to biaxial extension, the Alveograph (AlveoLAB, Chopin Technologies, Cedex, France) was used according to the AACC method 54.30-02. The optimum water absorption values determined in the Consistograph test were used (adapted hydration conditions). Doughs were prepared by mixing flour and whey with salted water and forming calibrated pieces of dough. After 20 min of resting time, the system inflates the test pieces to the point of rupture and records the pressure in the bubble as a function of time. In this model, temperature and hygrometry conditions are automated and fully controlled, and the inverted bubble is more spherical and closer to the ideal conditions of the test. Overpressure or dough´s tenacity (mmH_2O), extensibility (mm), P/L ratio and the work or deformation energy (10^{-4} J) were calculated. For each sample, five dough pieces were tested.

2.2.4. Small Amplitude Oscillatory Shear Measurements

SAOS measurements were performed in a controlled stress rheometer (Haake MARS III Thermo Fisher Scientific, Waltham, MA, USA) equipped with a UTC-Peltier and fitted with a serrated parallel plate system with 20 mm diameter and 2 mm gap. After kneading, the dough was shaped into small balls and fermented in the oven. The fermented samples were placed between the plate sensor and the dough surface exposed was coated with paraffin oil to prevent drying and allowed to rest 30 min before testing. Stress and frequency sweep tests were performed at 5 ± 1 °C to prevent fermentation during tests. The stress sweep test at 6.28 rad/s was always performed prior to the frequency sweep, for the determination of the linear viscoelastic zone. The viscoelastic properties of the dough were determined from the frequency sweep tests, applying a sinusoidally varying shear stress of 10 Pa over an angular frequency range of 0.001 to 100 rad/s. Two doughs of each formulation were prepared, separately, and tests were performed at least once in each sub-sample, corresponding to three repetitions in each dough sample.

2.2.5. Texture

Texture Profile Analysis

TPA was carried out in a temperature controlled room at 20 ± 1 °C, using a Texturometer TA.XTplus (Stable Micro Systems, Surrey, UK) equipped with a 5 kg load cell and a cylindrical acrylic probe of 10 mm diameter (p/10). Fermented doughs were placed in cylindrical containers with 25 mm height × 65 mm diameter and fermented before testing. TPA in penetration mode at 1 mm·s^{-1} of crosshead speed, 3 s of waiting time and 15 mm distance. The same samples prepared for SAOS measurements were used for texture, corresponding to three repetitions in each dough sample. Firmness (N), adhesiveness (N·s) and cohesiveness were the main representative parameters calculated from the texturograms.

SMS/Kieffer Dough and Gluten Extensibility Rig

For the uniaxial extensional tests, the SMS/Kieffer Dough and Gluten Extensibility Rig for the TA.XTplus was used, as described by Buresová et al. [28], with some modifications. The dough was molded into rolls and placed on the Teflon mold, forming test pieces with a length of 5 cm. The doughs were tested after resting for 10 min (t0) and 30 min (t30) at 30 °C. The force required to stretch the sample was recorded as a function of time using a test speed of 1.0 mm·s^{-1} and distance of 70 mm. The peak force—resistance to extension R (N), distance corresponding to this peak—extensibility E (mm), and ratio number R/E (N·mm^{-1}) are the most important parameters. The test was repeated at least four times for each sample.

Puncture Test

Bread crumb was measured 2 h after baking by means of a puncture test using a cylindrical acrylic probe of 19 mm diameter (p/19) at 2 mm·s^{-1} crosshead speed and 12 mm distance. Loaves were sliced by hand, 20 mm thick. Measurements were repeated four times for each sample, 2 h after baking (t0) and after two days of storage (t48 h). Firmness (N) was the texture parameter used to discriminate different bread samples.

2.2.6. Volume

Volume of the bread was measured using rapeseed displacement method AACC 10-05.01. In order to compare different breads, the same weight of ingredients, in relation to 300 g of wheat flour in mixture with whey powder, was used to prepare all the breads, using the formulations presented in Table 1. All the samples were evaluated in triplicate.

2.2.7. Statistical Analysis

The analysis of variance (one-way ANOVA) of the experimental data was performed using Origin Pro 8.0 software, followed by Tukey's test. Correlation analysis was performed by using STATISTICA (version 10.0). The significance level was set to 95% ($p < 0.05$).

3. Results and Discussion

3.1. Empirical Rheology of Dough

Figure 1 and Table 2 show the results obtained using empirical rheology equipment to characterize wheat doughs incorporated with different percentages of whey powder (0%—Control, 12%, 16% and 20% w/w). There was a decrease of the water absorption values obtained using Micro-doughLab and Consistograph when adding whey, decreasing dough development time (DDT), softening (DSO), and peak energy (PE), but increasing dough stability (DS), time to reach Prmax, tolerance to kneading (Tol) and dough consistency after 250 and 450 s.

Gélinas et al. [9] used fermented dairy products for bread and obtained the lowest water absorption value when whey (fermented or not) was used, and higher values of peak time and dough stability were measured in a Farinograph. In another study, the replacement of wheat flour with 5–15% whey powder concentrate also showed higher Farinograph stability and lower water absorption values [13]. Madenci and Bilgiçli [15] and Zhou et al. [17] also found a decrease in water absorption replacing wheat flour by whey protein, explaining that whey incorporation inhibited the hydration of granular starch and wheat proteins [17]. Zhou et al. [17] found that Mixolab dough stability time, an indicator of dough strength, decreased compared to the control for substitution levels of whey protein between 5% and 30%. They suggested that whey inhibits the gluten network structure due to the dilution effect of gluten and to the water competition between gluten, starch and whey. However, in several studies, the water absorption increased in the samples in which whey was added, and a negative impact on mixing properties of bread dough was observed [10,12].

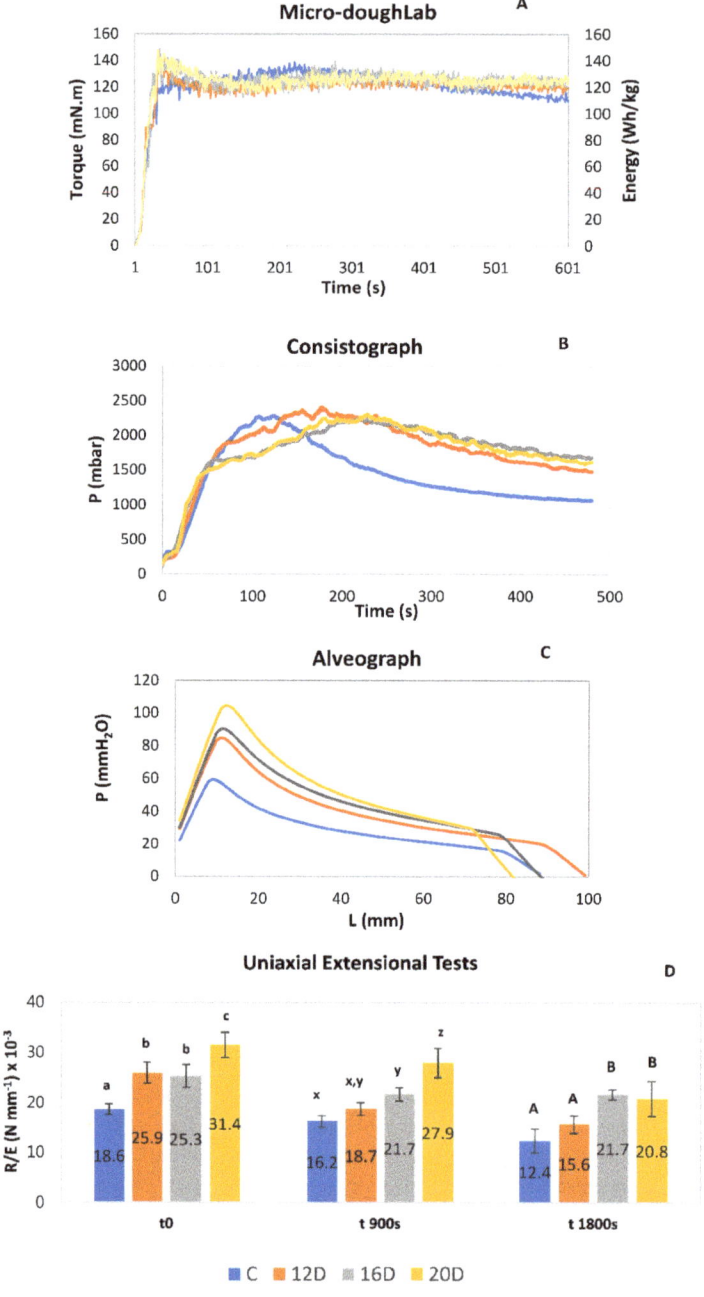

Figure 1. Curves obtained with Micro-doughLab (**A**), Consistograph (**B**) and Alveograph (**C**) and R/E values obtained from the extensional tests in the Texturometer (Kieffer dough rig) along fermentation (t0, t 900 s and t 1800 s) (**D**). C: control dough without whey; 12D: dough with 12% whey; 16D: dough with 16% whey; 20D: dough with 20% whey. Error bars indicate the standard deviations from the repetitions. Different letters (a, b, c or x, y, z or A, B) correspond to significant differences (one-way ANOVA, $p < 0.05$).

Table 2. Rheology parameters * obtained from Micro-doughLab mixture curves, Consistograph, Alveograph and Texturometer (Texture Profile Analysis and Extensibility tests). C: control dough without whey; 12D: dough with 12% whey; 16D: dough with 16% whey; 20D: dough with 20% whey. For each equipment, different letters (a, b, c) in the same column correspond to significant differences (one-way ANOVA, $p < 0.05$).

Micro-dougLab	WA (%)	P (mN.m)	DDT (s)	DS (s)	DSO (mN.m)	PE (Wh/kg)
C	52.2	130 ± 1.5	234 ± 0.3 [a]	258 ± 0.3 [a]	18.3 ± 0.6 [a]	15.0 ± 1.4 [a]
12D	39.4	128 ± 2.5	222 ± 2.3 [a]	558 ± 0.2 [b]	3.3 ± 1.5 [b]	15.5 ± 11.0 [a]
16D	38.2	133 ± 0.6	54 ± 0.1 [b]	552 ± 0.3 [b]	5.3 ± 1.1 [b]	2.9 ± 0.3 [b]
20D	36.4	133 ± 2.3	54 ± 0.1 [b]	522 ± 0.7 [b]	6.7 ± 1.5 [b]	3.0 ± 0.1 [b]

Consistograph	WA (%)	Prmax (mbar)	tPrmax (s)	Tol (s)	D250 (mbar)	D450 (mbar)
C	50.4	2204 ± 75	121 ± 4 [a]	129 ± 1 [a]	1403 ± 69 [a]	1262 ± 232 [a]
12D	39.6	2298 ± 2	171 ± 16 [b]	246 ± 20 [b]	2132 ± 87 [b]	1538 ± 61 [a,b]
16D	38.0	2203 ± 51	251 ± 20 [c]	307 ± 73 [b]	2194 ± 46 [b]	1755 ± 106 [b]
20D	36.2	2235 ± 64	235 ± 11 [c]	261 ± 19 [b]	2223 ± 72 [b]	1686 ± 94 [b]

Alveograph	WA (%)	P (mm H$_2$O)	L (mm)	P/L	W (10^{-4} J)	-
C	50.4	66.0 ± 1.2 [a]	78.4 ± 1.6 [a,b]	0.8 ± 0.1 [a]	158.7 ± 9.5 [a]	-
12D	39.6	92.7 ± 3.0 [b]	85.5 ± 1.1 [a]	1.1 ± 0.1 [a]	239.2 ± 12.5 [b]	-
16D	38.0	109.7 ± 13.7	71.6 ± 2.6 [b,c]	1.5 ± 0.4 [b]	258.3 ± 10.3 [b]	-
20D	36.2	109.8 ± 6.4	60.8 ± 2.5 [c]	1.8 ± 0.5 [b]	232.8 ± 13.7 [b]	-

Texturometer	WA (%)	Rmax t0 (N)	Emax t0 (mm)	Firmness (N)	Adhesiveness (N.s)	Cohesiveness
C	52.2	0.26 ± 0.029 [a]	13.93 ± 0.83 [a]	0.72 ± 0.30 [a]	6.89 ± 0.93 [a]	0.78 ± 0.16 [a]
12D	39.4	0.36 ± 0.054 [b]	13.83 ± 1.12 [a]	1.01 ± 0.13 [a]	8.63 ± 0.40 [a]	0.87 ± 0.10 [a]
16D	38.2	0.41 ± 0.029 [b,c]	16.38 ± 0.37 [b]	0.83 ± 0.09 [a]	8.63 ± 0.73 [a]	0.81 ± 0.04 [a]
20D	36.4	0.48 ± 0.037	15.16 ± 0.57 [a,b]	0.68 ± 0.07 [a]	5.59 ± 2.56 [a]	0.60 ± 0.14 [a]

* WA—water absorption. Micro-doughLab parameters: P–peak resistance; DDT—dough development time; DS—dough stability; DSO—dough softening; PE—peak energy. Consistograph parameters: Prmax—maximum pressure; tPrmax—time to reach Prmax; Tol—tolerance to kneading; D250 e D450—pressure after 250 and 450 s. Alveograph parameters: P—tenacity; L—extensibility; P/L—ratio between tenacity and extensibility; W—work or deformation energy. Extensibility tests: Rmax t0—resistance to extension before fermentation; Emax t0—extensibility before fermentation.

Concerning Alveograph results (Figure 1C and Table 2), one can see that doughs with higher levels of whey powder (16D and 20D) presented the highest tenacity and deformation energy and the lowest extensibility. From the P/L ratio values, it is possible to know about the elastic resistance of the dough to biaxial extension and the potential to produce bread [17,29]. When this value is within the range of 0.40 to 0.80, dough has balanced gluten, being suitable to produce breads, and P/L values lower than 0.4 indicates a very extendable dough. P/L values of the doughs enriched with high levels of whey (16D and 20D) were higher than 1.50, an indicator of a very strong dough, not suitable for bread production.

Regarding the extensibility tests in the Texturometer (Figure 1D and Table 2), an increase in dough resistance to extension (Rmax) with increasing whey powder concentration can be observed, while the extensibility before dough proofing (Emax) remained practically constant. The results for Rmax confirm the findings of previous studies carried out with whey ingredients, in which higher resistance values were obtained in an Extensograph in relation to control dough [12,13,15]. Erdogdu-Arnoczky et al. [10] found that the incorporation of whey solids did not affect the volume of CO_2 produced during proofing measured by a Rheofermentometer, but the dough ruptured earlier than the control, indicative of a weakened structure. The results of the present study show that whey addition exerted an effect on all doughs (before and after fermentation), increasing R/E modulus. This value is defined as the change of strain as a function of stress and is important in evaluating the balance between dough elasticity and extensibility. R/E varied from 18.6×10^{-3} N·mm^{-1} for control dough before fermentation (t0), typical of wheat, to 25.3 to 31.4×10^{-3} N·mm^{-1}, similar to the values reported by Buresová et al. [28] for corn, millet, quinoa and rice. However, when compared to the wheat control dough, the parameters obtained from the Texture Profile Analysis of the doughs with whey powder addition did not show significant differences ($p > 0.05$).

3.2. Fundamental Rheology of Dough

Fermented doughs have a viscoelastic behavior, with G′ higher than G″, and both values dependent on the frequency. A crossover of storage and loss moduli is observed at low frequencies (Figure 2). Based on G′ values at 1 Hz and 10 Hz, the addition of whey exerted a limited effect on the dynamic viscoelastic properties, since there were no significant differences ($p > 0.05$) between doughs with different composition. These results are consistent with the TPA parameters of the dough.

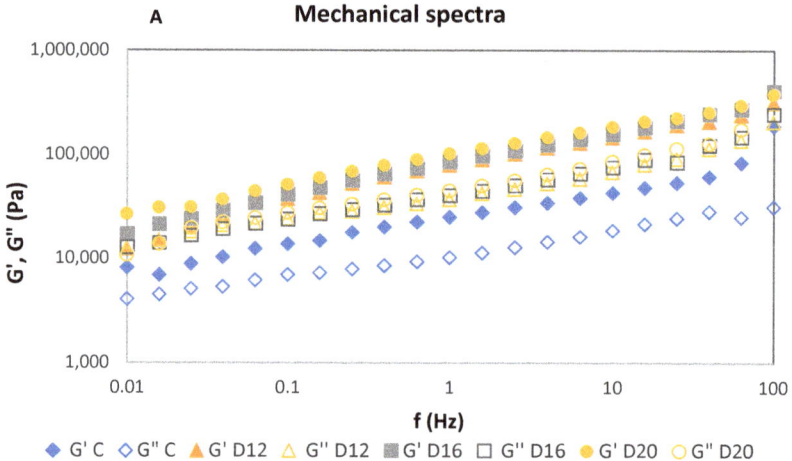

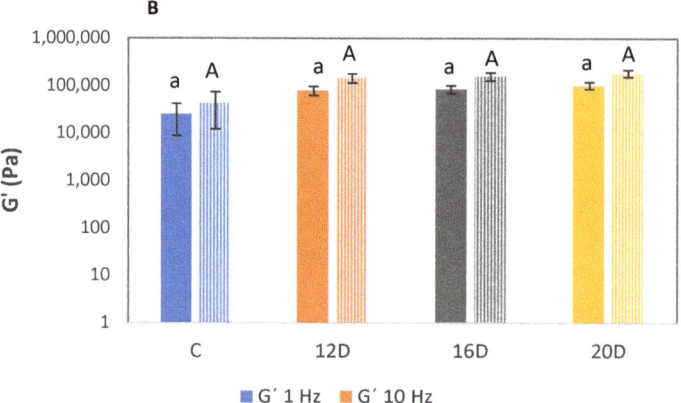

Figure 2. Mechanical spectra (**A**) and values of G′ at 1Hz obtained after dough fermentation (**B**). G′ (storage modulus—filled symbol), G″ (loss modulus—open symbol). C: control dough without whey; 12D: dough with 12% whey; 16D: dough with 16% whey; 20D: dough with 20% whey. Error bars indicate the standard deviations from the repetitions. Different letters (a, A) in the same graph correspond to significant differences (one-way ANOVA, $p < 0.05$).

McCann and Day [30], studying commercial wheat flours with different levels of protein, stated that the high phase volume of starch granules leads to the domination of starch–starch interactions over the protein phase. Therefore, the differences in the viscoelastic behavior of dough proteins are overshadowed by the high volume or starch, and changes in gluten structure are not accurately detected by the small deformation measurements. In addition, for wheat dough with whey protein, Zhou et al. [17] showed that viscoelastic moduli decreased and tanδ increased when increasing the whey protein proportion, indicating a weakened gluten network. This last study is in contradiction with the present findings and this should be due to different concentrations and experimental procedures used. In the present study, it is possible to observe a slight increase in the viscoelastic functions from control to the whey doughs; however, these were not statistically significant ($p < 0.05$). Therefore, we did not find a negative effect of the whey protein addition on the viscoelastic properties of dough, probably because the levels of whey protein addition were low and differences on G' and G" values were overshadowed by a high starch volume. Therefore, SAOS measurements were not accurate to detect differences between wheat doughs incorporated with different levels of the commercial whey powder. Contrary to the study mentioned [17], whey powder with only 12% of protein and not whey protein (over 76% protein) was used, and the difference in protein addition can explain these different results.

3.3. Breadmaking Properties

To study the relation between overall bread quality parameters and raw materials, the firmness and the volume of bread loaves were evaluated. For bread crumb differing in composition, crumb firmness increased with whey addition and there was a negative impact on loaf volume (Table 3).

Table 3. Values of crumb firmness and volume obtained for breads. C: control bread without whey; 12D: bread with 12% whey; 16D: bread with 16% whey; 20D: bread with 20% whey. Different letters (a, b, c, d) in the same column correspond to significant differences (one-way ANOVA, $p < 0.05$).

Sample	Firmness (N)		Volume (cm^3)
	t0	t48 h	
C	2.55 ± 0.39 [a]	4.68 ± 1.19 [a]	1272 ± 8 [a]
12D	4.51 ± 0.33 [b]	7.66 ± 1.93 [b]	1045 ± 9 [b]
16D	5.91 ± 0.34 [c]	6.89 ± 1.03 [a,b]	1152 ± 8 [c]
20D	4.14 ± 0.48 [b]	6.35 ± 0.78 [a,b]	958 ± 8 [d]

Erdogdu-Arnoczky et al. [10] reported that crumb softness and loaf volume increased with heat-treated acid whey protein addition. However, whey is known to have a negative effect in wheat bread loaves volume [12,13,16,31]. Zhou et al. [17] obtained a lower bread volume, increasing whey protein from 0% to 10%, but when whey content was higher than 20 %, bread volume was even higher than that of wheat bread control. They concluded that the level of the protein source plays an important role. For high whey levels, higher than gluten content, the heat induced whey protein gel became the dominant phase of the dough structure and gave strength to the expanding cells, resulting in higher loave volume. Wronkowska et al. [16] suggested the interaction between soluble whey proteins and the gluten proteins, weakening the elasticity of the gluten matrix structure. Moreover, a more rigid structure is obtained due to calcium, potassium and lactose, present in whey powder in high concentrations. According to the results in Table 3, it is possible to observe a trend on stiffening of the breads with the incorporation of whey, comparing to the control. However, it was not possible to establish a direct relationship between the volume reduction and the whey incorporation content.

There have been contradictory reports about the effect of whey on dough and bread quality, resulting from different interactions with wheat flour components. Those differences result from the different nature of the whey products varying from liquid or dried whole whey or whey protein concentrates or permeates, which are very different in composition. Furthermore, the baking method also affects the functionality of whey [15]. Therefore, it is crucial to study the functionality of the available whey products and their impact on the required applications in the bakery industry. Impact of different whey products on the technological performance of wheat dough is dependent on the whey level and composition.

3.4. Correlations between Parameters Obtained from Empirical and Fundamental Rheology

Using correlation analysis, the relationships between empirical and fundamental rheology properties of dough, with different amounts of whey powder incorporation were obtained. Only significant dependences ($r^2 > 0.60$) are presented in Table 4 and Figure 3.

Viscoelastic moduli of the doughs after proofing, obtained from fundamental SAOS measurements, are not correlated with the empirical rheology parameters and with breadmaking properties (firmness and volume). Other researchers studying wheat flour doughs suggested that the values for the dynamic moduli are not the most important factor determining bread making performance. There was no direct relationship between the dynamic moduli or tan δ and loaf volume [32].

Significant interdependences ($r^2 > 0.60$) within empirical rheology parameters of wheat dough samples were found (Table 4). Micro-doughLab parameters obtained from the mixture curves and Consistograph parameters are significantly correlated, and this relation was particularly relevant for dough stability (DS) vs. dough consistency after 250 and 450 s (D250 and D450). Some Micro-doughLab parameters, dough stability (DS) and softening (DSO), showed a significant positive correlation with dough tenacity (P) and energy used for the biaxial deformation (W) during the Alveograph test. Nevertheless, Micro-doughLab parameters are not correlated with texture parameters calculated from TPA and extensibility tests. As expected, considering that Micro-doughLab and Consistograph are both mixing devices, Consistograph parameters (tPrmax, D250 and D450) are also positively correlated with Alveograph P and W values. A positive relationship was also found between some Consistograph (tPrmax and D250) and Alveograph (P) parameters with the resistance to extension before dough fermentation (Rmax).

It was possible to establish correlations between empirical rheology dough properties and breadmaking performance. In Figure 3, one can see the bivariate scatterplots of significant ($r^2 > 0.60$) dependences—bread firmness vs. time to reach the maximum pressure (tPrmax), bread firmness vs. dough stability, bread firmness vs. dough softening, and bread volume vs. R/E modulus.

Matos and Rosell [31] found high correlation coefficients between dough Mixolab rheological parameters, namely dough consistency during mixing, and crumb hardness of rice-based gluten free bread. High extensional properties are a requisite for high wheat bread volumes, resulting in high crumb porosity, which causes a soft crumb texture [33]. In gluten free doughs, loaf volume was found to be in positive correlation with dough resistance to extension and dough extensibility under uniaxial deformation [28], in close agreement with the present results (Figure 3D).

Table 4. Correlations between dough rheology parameters obtained from empirical tests ($p < 0.05$).

Instrument	Micro-doughLab	Consistograph	Alveograph	Texturometer (Extensibility)
Micro-doughLab	-	DDT = 455 − 1.599 tPrmax ($r^2 = 0.71$) DS = 127 + 1.461 Tol ($r^2 = 0.69$) DS = −230 + 0.353 D250 ($r^2 = 0.87$) DS = −244 + 0.459 D450 ($r^2 = 0.87$) DSO = 24 − 0.067 Tol ($r^2 = 0.66$) DSO = 40 − 0.016 D250 ($r^2 = 0.61$) PE = 31 − 0.110 tPrmax ($r^2 = 0.65$)	DS = −42 + 5.418 P ($r^2 = 0.71$) DS = −63 + 2.377 W ($r^2 = 0.76$) DSO = 31 − 0.237 P ($r^2 = 0.61$) DSO = 33 − 0.108 W ($r^2 = 0.70$)	Not significant ($r^2 < 0.60$)
Consistograph	-	-	tPrmax = −34 + 2.405 P ($r^2 = 0.79$) D250 = 523 + 15.424 P ($r^2 = 0.77$) D250 = 567 + 6.320 W ($r^2 = 0.72$) D450 = 726 + 8.786 P ($r^2 = 0.65$)	tPrmax = 8 + 494 Rmaxt0 ($r^2 = 0.65$) D250 = 804 + 3137 Rmaxt0 ($r^2 = 0.62$) D250 = 576 + 56 Rmaxt0 ($r^2 = 0.63$)
Alveograph	-	-	-	P = 26 + 184 Rmaxt0 ($r^2 = 0.65$)
Texturometer (Extensibility)	-	-	-	-

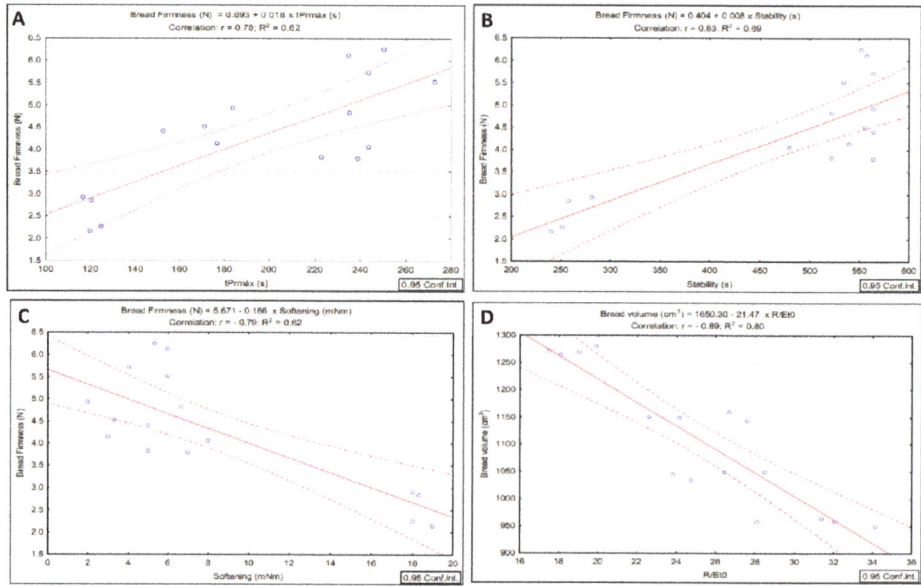

Figure 3. Correlations between breadmaking properties and dough rheology parameters ($p < 0.05$). (**A**): Bread firmness (Texturometer) vs. tPrmax (Consistograph); (**B**): Bread firmness (Texturometer) vs. Dough stability (Micro-doughLab); (**C**): Bread firmness (Texturometer) vs. Dough softening (Micro-doughLab); (**D**): Bread volume vs. R/E (t0) (extensional tests in the Texturometer).

4. Conclusions

In order to understand dough behavior during all breadmaking stages, different empirical rheological devices are used to imitate real processing conditions, which is very time consuming and requires a large amount of sample. The relationships between the parameters obtained by empirical devices and dough rheology and bread attributes can be useful to predict the behavior of the dough and provide important information for the bread making industry. Based on the findings of the present study, Micro-doughLab mixing parameters have a significant correlation with Consistograph and Alveograph data, and Consistograph with Alveograph values. The parameters obtained from Consistograph and Alveograph are significantly correlated with R/E extracted from extensibility tests using the Texturometer. Consistograph or Micro-doughLab devices can be used to estimate bread firmness and extensional tests of the dough may predict the volume of wheat loaves. Despite a negative impact on the empirical rheology parameters of the dough and poorer baking results, the use of this by-product should be considered for nutritional and sustainability reasons.

Author Contributions: C.M. was responsible for data acquisition, analysis and interpretation and for drafting the article. M.C.N. contributed to the data acquisition and interpretation and redaction of the manuscript. The study was designed and planned by I.S. and A.R., who revised the final version of the manuscript. All authors have read and agreed to the published version of the manuscript.

Funding: National funds from the Portuguese Foundation for Science and Technology (FCT) through the research unit UID/AGR/04129/2013-LEAF; PhD grant (Christine Macedo) from the University of Pará-Brazil.

Acknowledgments: This work was supported by Queijo Saloio S.A. Dairy Industry.

Conflicts of Interest: None of the authors have potential financial or other conflicts of interest to disclose.

References

1. Prazeres, A.; Carvalho, F.; Rivas, J. Cheese whey management: A review. *J. Environ. Manag.* **2012**, *110*, 48–68. [CrossRef] [PubMed]
2. Lappa, I.K.; Papadaki, A.; Kachrimanidou, V.; Terpou, A.; Koulougliotis, D.; Eriotou, E.; Kopsahelis, N. Cheese whey processing: Integrated biorefinery concepts and emerging food applications. *Foods* **2019**, *8*, 347. [CrossRef] [PubMed]
3. Madureira, A.; Pereira, C.; Gomes, A.; Pintado, M.; Malcata, F. Bovine whey proteins – overview on their main biological properties. *Food Res. Int.* **2007**, *40*, 1197–1210. [CrossRef]
4. Korhonen, H. Milk-derived bioactive peptides: From science to applications. *J. Funct. Foods* **2009**, *1*, 177–187. [CrossRef]
5. Kinsella, J.E.; Whitehead, D.M. Proteins in whey: Chemical, physical, and functional properties. *Adv. Food Nutr. Res.* **1989**, *33*, 343–438.
6. Fang, T.; Guo, M. Physicochemical, texture properties, and microstructure of yogurt using polymerized whey protein directly prepared from cheese whey as a thickening agent. *J. Dairy Sci.* **2019**, *102*, 7884–7894. [CrossRef]
7. Hadiyanto, H.; Ariyanti, D.; Aini, A.P.; Pinundi, D.S. Optimization of ethanol production from whey through fed-batch fermentation using *Kluyveromyces marxianus*. *Energy Procedia* **2014**, *47*, 108–112. [CrossRef]
8. Murari, C.S.; Machado, W.R.C.; Schuina, G.L.; Del Bianchi, V.L. Optimization of bioethanol production from cheese whey using *Kluyveromyces marxianus* URM 7404. *Biocatal. Agric. Biotechnol.* **2019**, *20*, 101182. [CrossRef]
9. Gélinas, P.; Audet, J.; Lachange, O.; Vachon, M. Fermented dairy ingredients for bread: Effects on dough rheology and bread characteristics. *Cereal Chem.* **1995**, *72*, 151–154.
10. Erdogdu-Arnoczky, N.; Czuchajowska, Z.; Pomeranz, Y. Functionality of whey and casein in fermentation and breadbaking by fixed and optimized procedures. *Cereal Chem.* **1996**, *73*, 309–316.
11. Kenny, S.; Wehrle, K.; Arendt, C.S.E.K. Incorporation of dairy ingredients into wheat bread: Effects on dough rheology and bread quality. *Eur. Food Res. Technol.* **2000**, *210*, 391–396. [CrossRef]
12. Bilgin, B.; Daglioglu, O.; Konyali, M. Functionality of bread made with pasteurized whey and/or buttermilk. *Ital. J. Sci.* **2006**, *3*, 277–286.
13. Indrani, D.; Prabhasankar, P.; Rajiv, J.; Rao, G.V. Influence of whey protein concentrate on the rheological characteristics of dough, microstructure and quality of unleavened flat bread (parotta). *Food Res. Int.* **2007**, *40*, 1254–1260. [CrossRef]
14. Asghar, A.; Anjum, F.M.; Allen, J.C.; Daubert, C.R.; Rasool, G. Effect of modified whey protein concentrates on empirical and fundamental dynamic mechanical properties of frozen dough. *Food Hydrocoll.* **2009**, *23*, 1687–1692. [CrossRef]
15. Madenci, A.B.; Bilgiçli, N. Effect of whey protein concentrate and buttermilk powders on rheological properties of dough and bread quality. *J. Food Qual.* **2014**, *37*, 117–124. [CrossRef]
16. Wronkowska, M.; Jadacka, M.; Soral-Smietana, M.; Zander, L.; Dajnowiec, F.; Banaszczyk, P.; Jelinski, T.; Szmatowicz, B. Acid whey concentrated by ultrafiltration a tool for modeling bread properties. *LWT Food Sci. Technol.* **2015**, *61*, 172–176. [CrossRef]
17. Zhou, J.; Liu, J.; Tang, X. Effects of whey and soy protein addition on bread rheological property of wheat flour. *J. Texture Stud.* **2018**, *49*, 38–46. [CrossRef] [PubMed]
18. van Riemsdijk, L.E.; van der Goot, A.J.; Hamer, R.J.; Boom, R.M. Preparation of gluten-free bread using meso-structured whey protein particle system. *J. Cereal Sci.* **2011**, *53*, 355–361. [CrossRef]
19. Kittisuban, P.; Ritthiruangdej, P.; Suphantharika, M. Optimization of hydroxypropylmethylcellulose, yeast b-glucan, and whey protein levels based on physical properties of gluten-free rice bread using response surface methodology. *LWT Food Sci. Technol.* **2014**, *57*, 738–748. [CrossRef]
20. Sahagún, M.; Gómez, M. Assessing influence of protein source on characteristics of gluten-free breads optimising their hydration level. *Food Bioprocess Technol.* **2018**, *11*, 1686–1694. [CrossRef]
21. Pico, J.; Reguilón, M.P.; Bernal, J.; Gómez, M. Effect of rice, pea, egg white and whey proteins on crust quality of rice flour-corn starch based gluten-free breads. *J. Cereal Sci.* **2019**, *86*, 92–101. [CrossRef]

22. Tomić, J.; Torbica, A.; Belović, M. Effect of non-gluten proteins and transglutaminase on dough rheological properties and quality of bread based on millet (Panicum miliaceum) flour. *LWT Food Sci. Technol.* **2020**, *118*, 108852. [CrossRef]
23. Graça, C.; Fradinho, P.; Sousa, I.; Raymundo, A. Impact of *Chlorella vulgaris* addition on rheology wheat dough properties. *LWT Food Sci. Technol.* **2018**, *89*, 466–474. [CrossRef]
24. Nunes, M.C.; Graça, C.; Vlaisavljevic, S.; Tenreiro, A.; Sousa, I.; Raymundo, A. Microalgae cell disruption: Effect on the bioactivity and rheology of wheat bread. *Algal Res.* **2020**, *45*, 101749. [CrossRef]
25. Dobraszczyk, B.J.; Morgenstern, M.P. Rheology and the breadmaking process. *J. Cereal Sci.* **2003**, *38*, 229–245. [CrossRef]
26. Song, Y.; Zeng, Q. Dynamic rheological properties of wheat flour dough and proteins. *Trends Food Sci. Technol.* **2007**, *18*, 132–138. [CrossRef]
27. Singh, S.; Singh, N. Relationship of polymeric proteins and empirical dough rheology with dynamic rheology of dough and gluten from different wheat varieties. *Food Hydrocoll.* **2013**, *33*, 342–348. [CrossRef]
28. Buresová, I.; Krácmar, S.; Dvoráková, P.; Streda, T. The relationship between rheological characteristics of gluten-free dough and the quality of biologically leavened bread. *J. Cereal Sci.* **2014**, *60*, 271–275. [CrossRef]
29. Al-Attabi, Z.H.; Merghani, T.M.; Ali, A.; Rahman, M.S. Effect of barley flour addition on the physico-chemical properties of dough and structure of bread. *J. Cereal Sci.* **2017**, *75*, 61–68. [CrossRef]
30. McCann, T.H.; Day, L. Effect of sodium chloride on gluten network formation, dough microstructure and rheology in relation to breadmaking. *J. Cereal Sci.* **2013**, *57*, 444–452. [CrossRef]
31. Matos, M.E.; Rosell, C.M. Quality indicators of rice-based gluten-free bread-like products: Relationships between dough rheology and quality characteristics. *Food Bioprocess Technol.* **2013**, *6*, 2331–2341. [CrossRef]
32. Janssen, A.M.; van Vliet, T.; Vereijken, J.M. Fundamental and empirical rheological behaviour of wheat flour doughs and comparison with bread making performance. *J. Cereal Sci.* **1996**, *23*, 43–54. [CrossRef]
33. Kieffer, R.; Wieser, H.; Henderson, M.H.; Graveland, A. Correlations of the breadmaking performance of wheat flour with rheological measurements on a micro-scale. *J. Cereal Sci.* **1998**, *27*, 53–60. [CrossRef]

© 2020 by the authors. Licensee MDPI, Basel, Switzerland. This article is an open access article distributed under the terms and conditions of the Creative Commons Attribution (CC BY) license (http://creativecommons.org/licenses/by/4.0/).

Article

On the Use of the Coaxial Cylinders Equivalence for the Measurement of Viscosity in Complex Non-Viscometric, Rotational Geometries

Regina Miriam Parlato [1], Eliana R. Russo [1], Jörg Läuger [2], Salvatore Costanzo [3], Veronica Vanzanella [3] and Nino Grizzuti [3,*]

1. Geolog Technologies, Viale Ortles 22/4, 20139 Milano, Italy; r.parlato@geolog.com (R.M.P.); e.russo@geolog.com (E.R.R.)
2. Anton Paar Germany GmbH, Helmuth-Hirth-Strasse 6, D-73760 Ostfildern, Germany; joerg.laeuger@anton-paar.com
3. Department of Chemical, Materials and Industrial Production Engineering (DICMaPI), University of Naples, P.le Tecchio 80, 80125 Naples, Italy; salvatore.costanzo@unina.it (S.C.); veronica.vanzanella@unina.it (V.V.)
* Correspondence: nino.grizzuti@unina.it

Received: 19 January 2020; Accepted: 27 March 2020; Published: 1 April 2020

Abstract: The rheology of macroscopic particle suspensions is relevant in many industrial applications, such as cement-based suspensions, synthetic and natural drilling fluids. Rheological measurements for these complex, heterogeneous systems are complicated by a double effect of particle size. On the one hand, the smallest characteristic length of the measuring geometry must be larger than the particle size. On the other hand, large particles are prone to sediment, thus calling for the use of rotational tools that are able to keep the suspension as homogeneous as possible. As a consequence, standard viscometric rotational rheometry cannot be used and complex flow geometries are to be implemented. In this way, however, the flow becomes non-viscometric, thus requiring the development of approximate methods to translate the torque vs. rotation speed raw data, which constitute the rheometer output, into viscosity vs. shear rate curves. In this work the Couette analogy methodology is used to establish the above equivalence in the case of two complex, commercial geometries, namely, a double helical ribbon tool and a square-shaped stirrer, which are recommended for the study of relatively large size suspensions. The methodology is based on the concept of the reduction of the complex geometry to an equivalent coaxial cylinder geometry, thus determining a quantitative correspondence between the non-standard situation and the well-known Couette-like conditions. The Couette analogy has been used first to determine the calibration constants of the non-standard geometry by using a Newtonian oil of known viscosity. The constants have been subsequently used to determine the viscosity curves of two non-Newtonian, shear thinning fluids, namely a homogeneous polymer solution and two heterogeneous concentrated suspensions. The results show that the procedure yields a good agreement between the viscosity curves obtained by the reduction method and those measured by a standard viscometric Couette geometry. The calibration constants obtained in this work from the coaxial cylinder analogy are also compared with those provided by the manufacturer, indicating that the calibration can improve the accuracy of the rheometer output.

Keywords: rheological measurements; non-viscometric geometries; Couette analogy; shear thinning fluids; suspensions

1. Introduction

Fluids used in the construction and Oil and Gas (O&G) industries share many common aspects. Typically, they are suspensions of solid particles in viscous matrices of various nature and are often characterized by very wide size and shape distributions. Examples include mortars and concrete in the construction field, and drilling fluids in the O&G industry [1,2]. The suspended solid phase determines a complex rheological behavior that, in turn, is not easy to quantify. Due to the broadness of the particle size distribution and to the (often large) density difference between the solid and the suspending liquid, particle settling is a major issue when rheological properties are to be measured [3]. In addition, the large particle size calls for equally large measuring geometries. Therefore, the classical, well-defined rotational rheometer geometries, such as the concentric cylinder geometry, produce wrong results. Therefore, one option for obtaining more accurate shear rate vs. shear stress data is to use unconventional, mixer-type geometries.

A mixer-type rheometer consists of an impeller with a relatively complex geometrical structure, rotating in a fluid contained in a tank, usually a cylindrical cup. The impeller shape and size are designed to minimize sedimentation effects and to allow for measurements on solid suspensions with large characteristic dimensions (typically, larger than one millimeter). This, however, makes the flow in the rheometer complex and non-viscometric. Consequently, both the shear stress and the shear rate, which are necessary to determine the fluid viscosity, are not defined. The instrument only returns a torque and a rotation speed value.

Many examples of mixer-type rheometers can be found in literature [4–9]. In particular, several authors described a procedure to convert the torque and rotational velocity data obtained from a vane-geometry rheometer into shear stress vs. shear rate relationships based on the use of the so-called Couette analogy [2,9–12]. Such a procedure consists in the reduction of the complex impeller geometry to an equivalent concentric cylinder geometry, whose dimensions are chosen in a way to match the viscosity vs. shear rate response of fluids of known rheology.

The main objective of the present work is to apply the above mentioned Couette analogy to the case of two mixer type geometries presently available for a commercial rotational rheometer [13], namely, a double helical ribbon tool and a square-shaped stirrer, indicated by the manufacturer as tools to measure the rheology of large-size concentrated suspensions [14]. The calibration constants for the two geometries are obtained by applying the Couette analogy to a Newtonian fluid and to a non-Newtonian, shear thinning polymer solution, whose rheological responses are quantitatively known. It is shown that the calibration constants provided by the manufacturer are different from those obtained in the present work, and that the latter give more accurate quantitative results for the non-Newtonian fluid. Finally, the two geometries are used to measure the rheology of two drilling fluids suspensions, showing that, when the Couette analogy is used to determine the correct calibration coefficients, both the double helical ribbon geometry and the square-shaped stirrer are able to accurately reproduce the non-Newtonian behavior of the fluids.

2. Theoretical Background

In rotational rheometry the so-called Couette, or Coaxial Cylinder (CC), geometry is widely used to determine the viscosity of fluids. Basically, the geometry consists of two coaxial cylinders, one still and the other rotating. Under these conditions, and assuming negligible inertial effects (low Reynolds number) the flow is viscometric, that is, a fluid element is always subjected to the same shear rate. As a consequence, by measuring the torque and the angular velocity at the moving cylinder under steady-state conditions, the viscosity of a Newtonian fluid can be measured [15].

Figure 1 refers to the CC geometry according to the ISO 3219:1995 Standard [16], showing also all the relevant geometrical parameters. In particular, for relatively narrow gaps, both shear rate and shear stress can be considered as uniform across the gap, thus allowing for the determination of the "true" or "representative" viscosity vs. shear rate curve also for non-Newtonian fluids [17]. The Narrow Gap Couette (NGC) geometry is today an international standard for the measurement of the non-Newtonian viscosity of polymers, emulsions and dispersions.

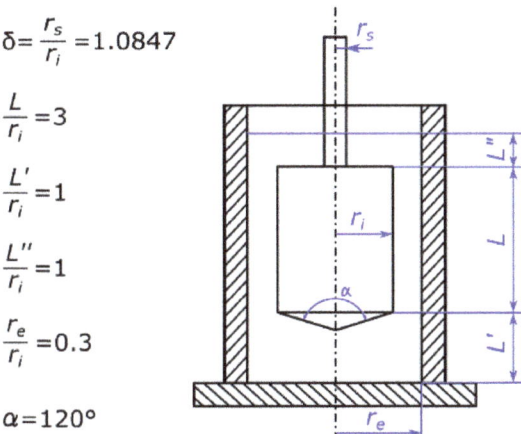

$$\delta = \frac{r_s}{r_i} = 1.0847$$

$$\frac{L}{r_i} = 3$$

$$\frac{L'}{r_i} = 1$$

$$\frac{L''}{r_i} = 1$$

$$\frac{r_e}{r_i} = 0.3$$

$$\alpha = 120°$$

Figure 1. The Concentric Cylinder (Couette) geometry and the geometry parameters according to the ISO 3219 Standard.

The NGC rheometry is governed by the following equations:

$$\sigma = \frac{1+\delta^2}{\delta^2} \frac{M}{4\pi L r_i^2 C_L} \qquad (1)$$

$$\dot{\gamma} = \frac{1+\delta^2}{\delta^2 - 1} \omega \qquad (2)$$

where σ and $\dot{\gamma}$ are the shear stress and the shear rate, respectively, $\delta = r_e/r_i$ is the ratio between the outer and inner radius, M and ω the measured torque and rotation speed, and L the height of the inner cylinder. C_L is an end-effect correction factor accounting for the torque acting at the end faces of the measuring system and is typically taken to be $C_L = 1.1$ [16].

All geometrical factors appearing in Equations (1) and (2) can be grouped into two constants, c_{SS} and c_{SR}:

$$c_{SS} = \frac{1+\delta^2}{\delta^2} \frac{1}{4\pi L r_i^2 C_L} \qquad (3)$$

$$c_{SR} = \frac{1+\delta^2}{\delta^2 - 1} \qquad (4)$$

Finally, the fluid viscosity, η, can be obtained as:

$$c_{SS} = \frac{\delta^2-1}{\delta^2} \frac{1}{4\pi L r_i^2 C_L} = \frac{c_{SS}}{c_{SR}} \frac{M}{\omega} = K \frac{M}{\omega} \qquad (5)$$

For a Newtonian fluid, for which the ratio between torque and rotation rate is a constant, K is the only geometrical constant to be known to determine the viscosity. In the non-Newtonian case, on the contrary, knowledge of c_{SS} and c_{SR} is required.

The well-defined situation of the NGC geometry finds no correspondence in the case of complex rotational geometries. Here, the flow field is a complex, non-uniform combination of shear and extensional components. The specific shape and size of the rotor, coupled to non-Newtonian characters of non-Newtonian fluid such as suspensions, make it impossible to determine the viscosity from simple formulas like Equation (5). In these cases, only numerical simulation techniques can in principle allow for a quantitative description of the flow characteristics. If, however, an approximate solution to the problem is sought, the so-called "Couette analogy" approach can be followed [9,11]. The approach is based on the simple idea that the complex geometry can be *reduced* to a virtual coaxial cylinder geometry such that, for a given rotation rate, the complex geometry and its CC virtual equivalent produce the same torque.

In its simplest form [9], the Couette analogy assumes that the fluid is Newtonian and that the geometry consists of an outer cylindrical cup, of radius r_e, where the rotor is immersed in the fluid up to a given height L. The complexity of the geometry resides in the inner rotating tool of non-cylindrical shape. Let now ω and M be the rotation rate and the torque measured by the complex geometry on a fluid of known viscosity, η. According to the Couette analogy, it is assumed that the non-viscometric flow is equivalent to that taking place in a virtual coaxial cylinder geometry with the same r_e and L and of unknown inner radius $r_{i,eq}$. The latter can be found by rewriting Equation (5) as:

$$4\pi L C_L \eta_{eq} \frac{\omega}{M} = \frac{r_e^2 - r_{i,eq}^2}{r_e^2 r_{eq}^2} \qquad (6)$$

so that the inner radius of the equivalent Couette geometry is given by:

$$r_{i,eq} = \frac{r_e}{\sqrt{1 + 4\pi L C_L \eta_{eq} r_e^2 \frac{\omega}{M}}} = \frac{r_e}{\sqrt{1 + 4\pi L C_L \eta_{eq} K r_e^2}} \qquad (7)$$

In the second equality of Equation (7) use has been made of the definition of the calibration constant K. This proves that, in order to determine the Couette analogue of the complex geometry for the case of a Newtonian fluid, only the calibration constant of the non-viscometric geometry is required.

The reduction of a complex rotational geometry to its Couette equivalent holds in principle only for the case of a Newtonian fluid. Attempts have been made in the literature to include the non-Newtonian constitutive behavior into the analogy [9,12]. However, adding the non-Newtonian complexity to a model that is already based on very crude assumptions is not particularly meaningful. For this reason, in order to validate the model for the complex geometries used here, in the next section the experimental results for both Newtonian and non-Newtonian fluids are compared by keeping the Couette analogy at its simplest, yet more affordable and robust level described above.

3. Materials and Methods

Four different fluids were used to test the non-viscometric geometries. They are:

(1) A Newtonian silicon oil (BDH1000, from Merck KGaA, Darmstadt, Germany) with a viscosity of about 1 Pa s (labeled SO);
(2) A shear thinning, homogeneous aqueous solution of HydroxyEthyl Methyl Cellulose (TyloseMH600046P6, manufactured by SE Tylose GmbH, Wiesbaden, Germany). The concentration is 0.7% wt, which determines (see data of Figure 7) a Newtonian plateau viscosity of about 0.7 Pa·s (labeled HEMC);
(3) Two commercial drilling fluids provided by Geolog Srl (Milan, Italy). They are a water-based (labeled WBM) and an oil-based (OBM) suspension containing sand particles of variable size up to about 1 mm and a volume fraction of about 10%. The suspending fluid contains several additives, including polymers, but their detailed formulation is not known for confidential reasons. Both suspensions show a strong shear thinning behavior over a wide range of shear rates and no hint of a low shear Newtonian plateau.

All fluids were tested by an Anton Paar MCR702 rotational rheometer (Anton Paar GmbH, Graz, Austria). The rheometer is equipped with Peltier units for an accurate thermal control. In all experiments, temperature has been kept constant at 25 °C. Measurements were repeated at least four times for each fluid and each geometry. In all cases, a very good reproducibility was obtained. For this reason, only single run experiments are shown.

Three different rotational geometries have been used, manufactured by Anton Paar for the MCR rheometer line. Pictures and drawings of the rotors are given in Figures 2 and 3. They are:

(1) A Narrow Gap Couette geometry (NGC) formed by an inner cylinder (Figure 2a) and an outer cup (Figure 2b). The geometry conforms to the ISO 3219 Standard and has been used to calibrate the other geometries via the Couette analogy;
(2) A Double Helix (DX) rotor (Figure 2c) that uses the same cup of the NGC. The shape of the DH rotor introduces a continuous flow from the bottom to the top of the cell, thus opposing particle sedimentation;
(3) A flat, square-shaped blade rotor (Building Material Cell, BMC, Figure 2c–e) fitted in a cylindrical cup larger than that used by the NGC and DH configurations. The cup includes an inner cage to prevent slippage of the fluid. It must be noticed that, according to the manufacturer, the tool can be used to measure suspensions having aggregates no larger than 5 mm. Other rotors are available for larger particle suspensions.

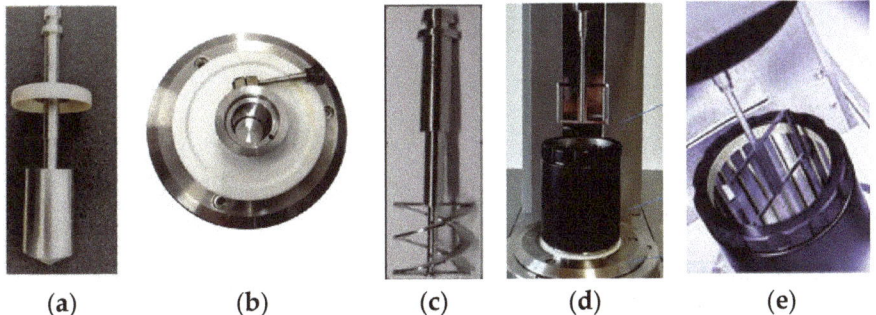

Figure 2. The rotors used in this work: (**a**) Narrow Gap Couette (ISO 3219); (**b**) Top view of the cup used for both the Narrow Gap Couette (NGC) and the Double Helix geometry; (**c**) Double Helix (DH); (**d**) Building Material Cell (BMC); (**e**) A close-up of the BMC cup.

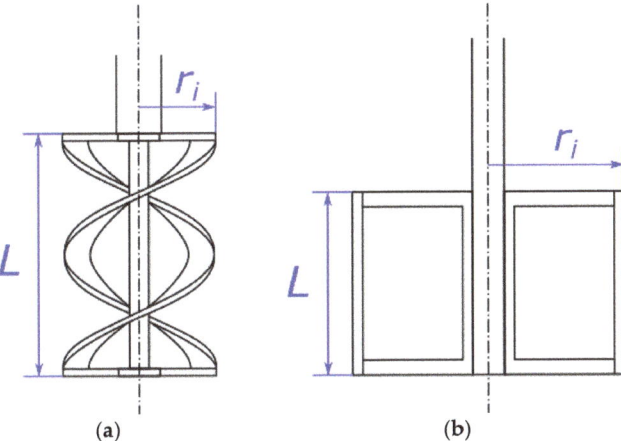

Figure 3. Drawing of the two complex rotors and their main dimensions: (**a**) Double Helix (DH); (**b**) Building Material Cell (BMC).

Drawings of the DH and BMC rotors are shown in Figure 3 along with their main geometrical dimensions (the same drawing for the NGC has been already reported in the previous Figure 1). Dimensions of inner and outer radii and rotors height are summarized in Table 1.

Table 1. Main geometrical parameters of the three geometries. The equivalent internal radius, obtained from the Couette analogy procedure described in the text, is also reported for the two non-conventional geometries.

Geometry	r_e (mm)	r_i (mm)	$r_{i,eq}$ (mm)	L (mm)
NGC	14.46	13.33	/	39.98
DH	14.46	12.00	10.446	37.00
BMC	35.0	29.5	23.358	44.3

4. Experimental Results and Discussion

4.1. Calibration of Non-Conventional Geometries

The Newtonian silicon oil has been used to calibrate the complex geometries and to determine the corresponding coefficients. To this end, the NGC geometry has been used as a reference. Figure 4 shows the steady-state torque as a function of the rotation speed for the three rotors.

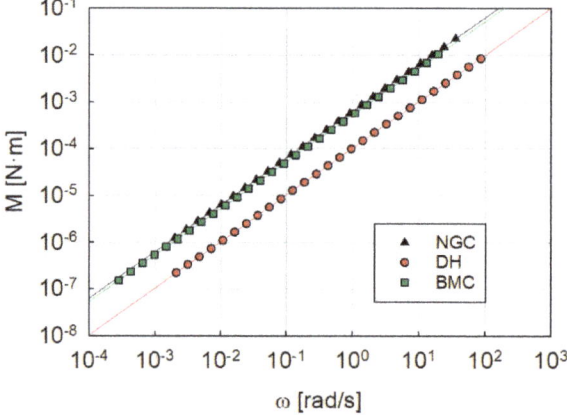

Figure 4. Torque as a function of rotation rate for the three geometries. The straight solid lines are linear regressions of the data.

For all three geometries the log-log data are well described by straight regression lines of slopes equal to one (within three significant digits) over about five decades of rotation rate and torque, indicating an almost perfect linearity. This is an expected result, in light of the Newtonian character of the SO fluid.

The data in Figure 4 can now be converted into viscosity vs. shear rate curves. To do that, Anton Paar provides the constants c_{SS} and c_{SR} for the three geometries. They are listed in Table 2. While the constants for the NGC geometry are rigorously determined from Equations (3) and (4), those provided for the DH and BMC rotors are empirical, although no details of the calibration procedure are given.

Table 2. The geometrical constants c_{SS} and c_{SR} for the three geometries. The values of c_{SS} and c_{SR} provided by Anton Paar are also reported along with those obtained in this work. The calibration constants K for the DH and the BMC derived from the calibration procedure are also reported.

Geometry	c_{SS} (1/m^3)		c_{SR} (1/rad)		K (rad/m^3)
-	Anton Paar	Couette Analogy	Anton Paar	Couette Analogy	-
NGC	18,847	-	12.337	-	-
DH	90,700	39,420	9.5493	4.1834	9422.2
BMC	14,070	6584.8	9.5493	3.6061	1826.0

Figure 5a shows the viscosity of the silicon oil as a function of shear rate for the NGC and the two complex geometries, obtained by using the standard Anton Paar geometrical coefficients. It is apparent that the three geometries return different values of the viscosity. The viscosity measured by the NGC tool corresponds to the value provided by the oil manufacturer. In particular, the NGC geometry returns an average viscosity $\eta = 0.9604$ Pa·s. The viscosity obtained from the DH and the BMC substantially differ from each other. In particular, the viscosity of the DH system matches the true value obtained by the NGC geometry, whereas the BMC system underestimates the viscosity by about 20%. This means that, at least for the BMC rotor, the coefficients c_{SS} and c_{SR} provided by the rheometer manufacturer are not accurate.

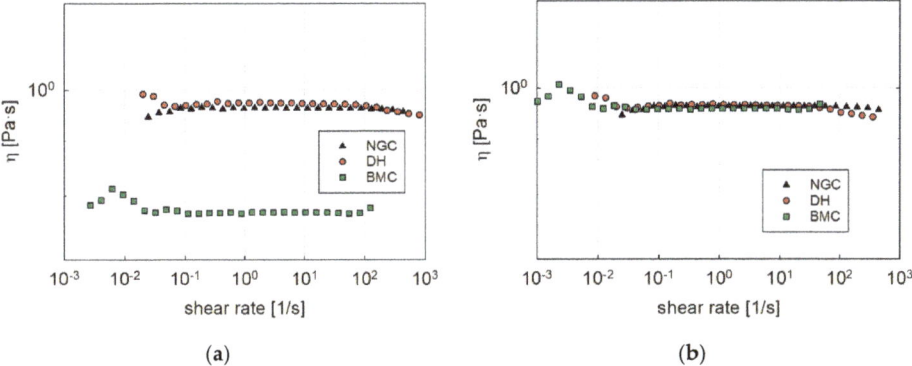

Figure 5. The viscosity of the silicon oil (SO) fluid as a function of shear rate for the NGC and the two complex geometries. (**a**) Results obtained by using the c_{SS} and c_{SR} coefficients provided by Anton Paar; (**b**) Results obtained by using the c_{SS} and c_{SR} coefficients from the Couette analogy.

Having verified that the NGC geometry returns the true viscosity value, the latter has been used to determine the new c_{SS} and c_{SR} coefficients for the DH and BMC tools based on the Couette analogy procedure described in Section 2. First, for each geometry the calibration constant K has been calculated from equation (5). To this end, the average viscosity measured with the NGC geometry is used, along with the slope of the torque vs. rotation rate linear plot (see Figure 4). Then, the Couette analogy is applied and the inner radius of the virtual concentric cylinder analogue, $r_{i,eq}$, is calculated by using Equation (7), where the necessary geometrical parameters are those listed in Table 1. Finally, use of Equations (3) and (4) allows for the determination of the two coefficients c_{SS} and c_{SR}.

The numerical values resulting from the above procedure are listed in Table 1 ($r_{i,eq}$) and Table 2 (K, c_{SS} and c_{SR}). It can be noticed that the coefficients calculated by using the Couette analogy are substantially different from those provided by Anton Paar. Figure 5b shows the viscosity curves obtained from the Couette analogy. All experimental data, including those obtained with the BMC geometry, now superimpose on those measured by the NGC geometry. This is an obvious result, as the analogy is based on the assumption that the correct viscosity value is the one measured by the NGC geometry, which is then used to "constrain" the coefficients of the other geometries to return the correct viscosity value. Notice, however, that the experimental data for the two non-viscometric geometries are shifted along the shear rate axis, due to the change in the c_{SR} coefficient.

4.2. Validation of the Couette Analogy Calibration Procedure

Once the c_{SS} and c_{SR} coefficients for each geometry are determined by the Couette analogy in a way to predict the correct Newtonian viscosity, they can be used to compare the NGC measurements performed on non-Newtonian fluids with those of the non-viscometric, complex rotors.

The results of the viscosity measurements for the shear thinning, HEMC solutions are reported in Figure 6. As in the previous Figure 5, the viscosity vs. shear rate curves obtained by using the geometry coefficients provided by Anton Paar are compared with those calculated from the Couette analogy.

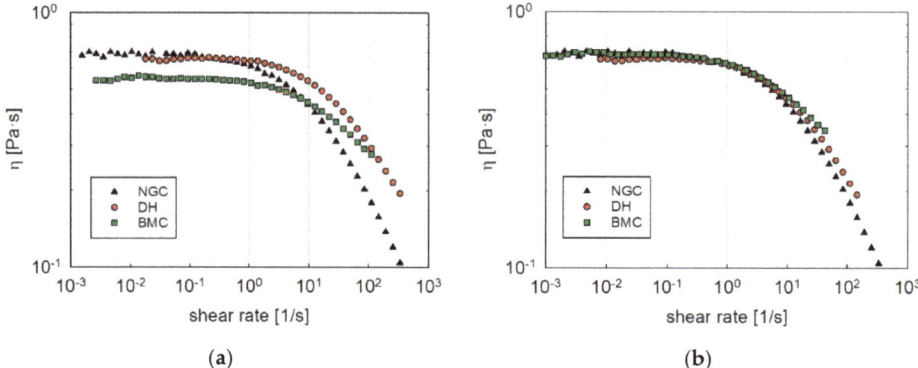

Figure 6. The viscosity of the HydroxyEthyl Methyl Cellulose (HEMC) aqueous solution as a function of shear rate for the NGC and the two complex geometries. (**a**) Results obtained by using the c_{SS} and c_{SR} coefficients provided by Anton Paar; (**b**) Results obtained by using the c_{SS} and c_{SR} coefficients from the Couette analogy.

When the Anton Paar constants are used (Figure 6a), substantial discrepancies are found between the NGC results and those obtained with the complex rotational geometries. In particular, the BMC impeller shows a Newtonian plateau viscosity lower than that measured with the NGC geometry, with a behavior similar to the one already observed for the silicon oil. The relative error is also in this case of the order of 20%. At higher shear rates, on the contrary, the viscosity from the BMC geometry becomes larger than that measured with the NGC, a cross-over between the two curves taking place at a shear rate of ca. $1\,\text{s}^{-1}$. As far as the DH geometry is concerned, the Newtonian plateau viscosity is well predicted as for the silicon oil measurements, but the shear thinning part of the curve is shifted to higher shear rates with respect to the reference NGC geometry data.

The results improve considerably when the c_{SS} and c_{SR} coefficients coming from the Couette analogy are used, as shown in Figure 6b. In this case, the viscosity curves obtained with the DH and BMC rotors match very closely the true viscosity data obtained by the NGC geometry. The agreement is particularly good for the Double Helix, whereas the deviation is larger for the BMC impeller. The DH data clearly indicate that the main reason for measurement improvement derives mostly from the horizontal shift in the data when the Couette analogy is used. This proves that the analogy allows for a robust estimate of the characteristic velocity gradient taking place in these non-viscometric, complex flow geometries.

The data presented in Figure 6 confirm that, even for a shear thinning homogeneous fluid, the Couette analogy provides an excellent interpretation of the experimental data. The situation becomes more challenging in the case of the non-Newtonian drilling fluid suspensions. Figure 7 reports the viscosity as a function of shear rate for the WBM suspension. In this case, where a non-Newtonian plateau is not present, at least in the shear rate range explored, both the DH and BMC viscosity qualitatively follow the NGC geometry behavior when the factory coefficients are used (Figure 7a). The agreement becomes remarkable when the values of c_{SS} and c_{SR} are calculated by the Couette analogy procedure (Figure 7b).

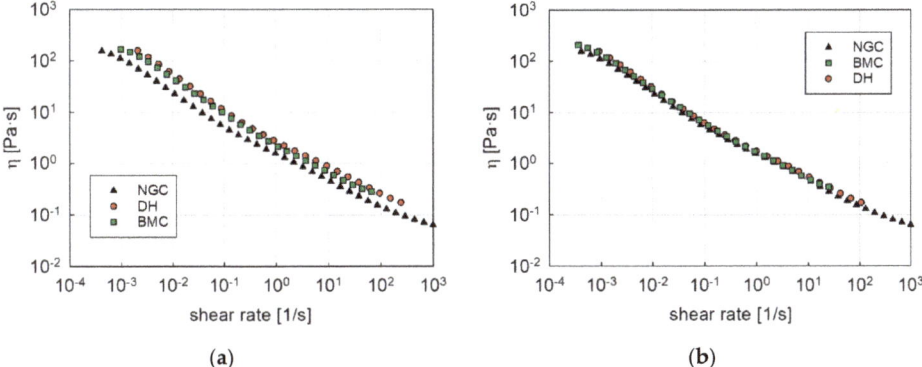

Figure 7. The viscosity of the WBM drilling fluid suspension as a function of shear rate for the NGC and the two complex geometries. (**a**) Results obtained by using the c_{SS} and c_{SR} coefficients provided by Anton Paar; (**b**) Results obtained by using the c_{SS} and c_{SR} coefficients from the Couette analogy.

Figure 8 shows the last set of viscosity measurements, involving the OBM suspension. In this case the experimental results are more contradictory. In the high shear rate region, above about 1 s^{-1}, the agreement between the three geometries is qualitatively good when the Anton Paar coefficients are used and becomes quantitatively excellent when the Couette analogy is implemented. On the other hand, when the low shear rate region is considered, the NGC data strongly differ from those obtained from the DH and BMC geometries. The discrepancy, in this case, is probably due to an incorrect evaluation of the viscosity in the NGC tool. One possible explanation is that, due to the non-polar character of the suspending fluid (oil) with respect to the electrically active sand particles, the system is prone to formation of particle aggregates at low rotation rates. Such aggregates are excluded from the relatively narrow gap measuring region of the concentric cylinders, thus determining an apparent lower viscosity of the suspension. Conversely, at higher shear rates the agglomerates are probably destroyed by the action of flow, thus restoring the well-dispersed suspension condition and, as a consequence, the correct viscosity behavior.

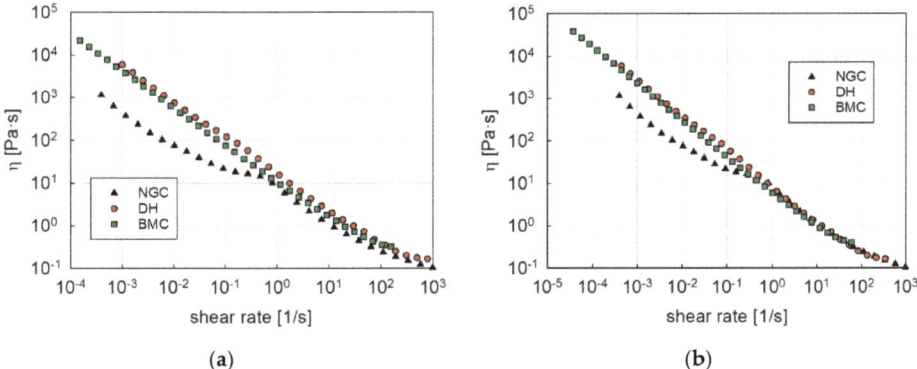

Figure 8. The viscosity of the OBM drilling fluid suspension as a function of shear rate for the NGC and the two complex geometries. (**a**) Results obtained by using the c_{SS} and c_{SR} coefficients provided by Anton Paar; (**b**) Results obtained by using the c_{SS} and c_{SR} coefficients from the Couette analogy.

5. Conclusions

The main conclusion of this work is that the Couette analogy concept, which allows for the reduction of a complex, non-viscometric rotational geometry to a virtual concentric cylinder analogue, can be successfully applied to determine the viscosity vs. shear rate curve of non-Newtonian fluids in complex geometries. The methodology has been applied to two commercial rheological tools, namely, a Double Helix (DH) and a flat square-shaped (BMC) rotor. It has been found that the standard geometrical coefficients provided by Anton Paar to convert the torque vs. rotation rate measurements into viscosity vs. shear rate curves do not produce accurate predictions. On the contrary, once the Couette analogy is implemented by using a Newtonian fluid as a rheological reference, the two complex geometries are able to reproduce, to a great level of accuracy, the non-Newtonian behavior of homogeneous as well as heterogeneous, suspension-based fluids. It must be also remarked that, although the non-conventional geometries are characterized by a complex flow field, including both shear and extensional non uniform components, the Couette analogy procedure allows for an accurate description of the viscous behavior of complex fluids over a very wide range of shear rates, sometimes extending over as many as eight decades.

Author Contributions: Conceptualization and methodology, N.G.; formal analysis, N.G. and J.L.; investigation, R.M.P., S.C. and V.V.; resources, E.R.R.; data curation, S.C. and N.G.; writing—original draft preparation, R.M.P.; writing—review and editing, N.G.; supervision, E.R.R. and N.G. All authors have read and agreed to the published version of the manuscript.

Funding: This research received no external funding.

Acknowledgments: We are grateful to Michela Brunelli for performing some of the rheological measurements and to Anton Paar GmbH for providing all technical specifications of the geometries tested in this work.

Conflicts of Interest: R.M.P. and E.R.R. are employees of Geolog. J.L. is an employee of Anton Paar. All of them state that in this paper there is nothing that may be considered as a conflict of interest.

References

1. Roussel, N. *Understanding the Rheology of Concrete*; Elsevier Ltd.: Amsterdam, The Netherlands, 2011; ISBN 978-0-85709-028-7.
2. Coussot, P. *Mudflow Rheology and Dynamics*; Balkema, A.A., Ed.; IAHR/AIRH Monograph Series; CRC Press: Boca Raton, FL, USA, 1997; ISBN 90-5410-693-X.
3. Chien, S.-F. Settling velocity of irregularly shaped particles. *Spe Drill. Completion* **1994**, *9*, 281–289. [CrossRef]
4. Hu, C.; De Larrard, F.; Sedran, T.; Boulay, C.; Bosc, F.; Deflorenne, F. Validation of BTRHEOM, the new rheometer for soft-to-fluid concrete. *Mater. Struct. Mater. Constr.* **1996**, *29*, 620–631. [CrossRef]
5. Wallevik, J.E. Minimizing end-effects in the coaxial cylinders viscometer: Viscoplastic flow inside the ConTec BML Viscometer 3. *J. Non-Newton. Fluid Mech.* **2008**, *155*, 116–123. [CrossRef]
6. Tattersall, G.H.; Banfill, P.F.G. Relationships between the British Standard Tests for Workability and the Two-Point Test. *Mag. Concr. Res.* **1977**, *29*, 156–158. [CrossRef]
7. Beaupré, D.; Lacombe, P.; Khayat, K.H. Laboratory investigation of rheological properties and scaling resistance of air entrained self-consolidating concrete. *Mater. Struct. Mater. Constr.* **1999**, *32*, 235–240. [CrossRef]
8. Koehler, E.P.; Fowler, D.W. Comparison of workability test methods for self-consolidating concrete. *J. ASTM Int.* **2010**, *7*, 1–19. [CrossRef]
9. Aït-Kadi, A.; Marchal, P.; Choplin, L.; Chrissemant, A.-S.; Bousmina, M. Quantitative analysis of mixer-type rheometers using the couette analogy. *Can. J. Chem. Eng.* **2002**, *80*, 1166–1174. [CrossRef]
10. Estellé, P.; Lanos, C.; Perrot, A.; Amziane, S. Processing the vane shear flow data from Couette analogy. *Appl. Rheol.* **2008**, *18*, 1–6. [CrossRef]
11. Lacoste, C.; Choplin, L.; Cassagnau, P.; Michel, A. Rheology innovation in the study of mixing conditions of polymer blends during chemical reaction. *Appl. Rheol.* **2005**, *15*, 314–325. [CrossRef]
12. Paruta-Tuarez, E.; Marchal, P.; Choplin, L. Application of the systemic rheology to the in situ follow-up of viscosity evolution with reaction time in the synthesis of urethane prepolymers. *Int. J. Adhes. Adhes.* **2014**, *50*, 32–36. [CrossRef]
13. Qin, Z.; Suckale, J. Flow-to-sliding transition in crystal-bearing magma. *J. Geophys. Res. Solid Earth* **2020**, *125*, e2019JB018549. [CrossRef]
14. Anton Paar GmbH. Available online: http://www.anton-paar.com (accessed on 27 March 2020).

15. Mezger, T.G. *The Rheology Handbook: For Users of Rotational and Oscillatory Rheometers*; Coatings Compendia; Vincentz Network: Hannover, Germany, 2006; ISBN 978-3-87870-174-3.
16. *ISO 3219:1995 Plastics—Polymers/Resins in the Liquid State or as Emulsions or Dispersions—Determination of Viscosity Using a Rotational Viscometer with Defined Shear State*; ISO: Geneva, Switzerland, 1995; ISBN 978-0-12-409547-2.
17. Grizzuti, N. Rheometry. In *Reference Module in Chemistry, Molecular Sciences and Chemical Engineering*; Elsevier: Amsterdam, The Netherlands, 2014; ISBN 978-0-12-409547-2.

 © 2020 by the authors. Licensee MDPI, Basel, Switzerland. This article is an open access article distributed under the terms and conditions of the Creative Commons Attribution (CC BY) license (http://creativecommons.org/licenses/by/4.0/).

Article

An Experimental Study on Human Milk Rheology: Behavior Changes from External Factors

Diana Alatalo * and Fatemeh Hassanipour *

Department of Mechanical Engineering, The University of Texas at Dallas, 800 W. Campbell Rd, Richardson, TX 75080, USA
* Correspondence: diana.alatalo@utdallas.edu (D.A); fatemeh@utdallas.edu (F.H.)

Received: 29 February 2020; Accepted: 18 March 2020; Published: 27 March 2020

Abstract: The influence of external factors, including temperature, storage, aging, time, and shear rate, on the general rheological behavior of raw human milk is investigated. Rotational and oscillatory experiments were performed. Human milk showed non-Newtonian, shear-thinning, thixotropic behavior with both yield and flow stresses. Storage and aging increased milk density and decreased viscosity. In general, increases in temperature lowered density and viscosity with periods of inconsistent behavior noted between 6–16 °C and over 40 °C. Non-homogeneous breakdown between the yield and flow stresses was found which, when coupled with thixotropy, helps identify the source of nutrient losses during tube feeding.

Keywords: human milk; tube feeding; breastfeeding; viscosity; complex modulus; density

1. Introduction

Milk is a species-specific bio-fluid produced in the mammary gland and traditionally fed directly to young at the breast. In humans, exclusive breastfeeding is recommended for the first 6 months of life with continued breastfeeding until 1 year of age or longer with introduction of complementary foods [1]. When infants cannot feed directly at the breast, mothers express their milk to be fed by artificial methods, such as gastric tubes, cups, or bottles [2]. Storage and feeding methods can result in nutrient losses in expressed human milk [3–5] which indicates that rheological changes in milk occur during storage and negatively impact flow. The main aim of this work is to explore rheological behavior of human milk and how external factors impact that behavior.

Most rheological studies on mammalian milk occur with dairy animals, predominately bovine. Studies often formed relationships between specific components in milk (i.e., fat), viscosity, and some external factor. External factors known to impact viscosity include pressure, temperature, pH, pasteurization, and homogenization [6–17]. Regression equations have been developed for different applications and are presented in Table 1. These equations show that the basic rheological behavior of bovine milk in regards to temperature appears to be consistent; milk viscosity decreases as temperature increases. This decrease is dependent on whether or not the proteins have begun to denature [10] and the concentration of proteins and fats. Difficulty comparing the studies lies in the lack of details, which are summarized in Appendix A.

Modern studies use homogenized and pasteurized milk or milk products and assume Newtonian behavior. However, an early work [6] found that raw bovine milk, both skim and whole, is a shear-thinning non-Newtonian fluid with greater variability at low pressures. Repeated trials of milk through capillary tubes at constant pressure resulted in lowering whole milk viscosity with repeated runs (thixotropic) but no change in skim milk viscosity. The decrease in whole milk viscosity with repeated experiments was even more evident as the age of the milk increased. Repeated trials with homogenized whole milk produced results similar to skim milk with viscosity remaining constant during each run. The authors attributed the decrease in viscosity in raw whole milk from repeated runs through a capillary to clumps of fat globules breaking up. They further investigated the effect of aging on raw skim milk viscosity since previous studies found aging to increase the viscosity of raw whole milk. Their results found that refrigeration of skim milk increased viscosity but freezing initially decreased viscosity which later increased with longer freeze times. The authors concluded that bovine milk viscosity was dependent on shearing force, age, method of storage, and mechanical agitation (for raw whole milk). All testing occurred at 25 °C, below the melting temperature of bovine milk fat [18].

Table 1. Regression equations for bovine viscosity and density.

Author	Regression Equation and Nomenclature
Snoeren et al. [10]	$\mu = \mu_{ref}[1 + \frac{1.25(\phi_c + \phi_{nw} + \phi_{dw})}{1 - \frac{\phi_c + \phi_{nw} + \phi_{dw}}{\phi_{max}}}]^2$ $\mu_{ref}(cP)$: viscosity of medium ϕ_c: volume fractions of casein ϕ_{nw}: volume fractions of native whey protein ϕ_{dw}: volume fractions of denatured whey protein ϕ_{max}: maximum volume fractions of all protein
Jebson and Chen [11]	$\ln \mu = 3.911 + 0.0202(\frac{S - 482.5}{0.85}) - 0.1291(\frac{T - 52.5}{7.5})$ S: solids content (g/kg) T: temperature (°K)
Phipps [13]	$\log_{10} \mu = [1.2876 + 11.07 \times 10^{-4} T_C][F + F^{\frac{5}{3}}] + \frac{0.7687 \times 10^3}{T_K} - 2.4370$ F: fat content (%) T_C: temperature (°C) T_K: temperature (°K)
Bakshi and Smith [12]	$\ln \mu = -8.9 + 0.1F + \frac{2721.5}{T}$ $\rho = 0.3T - 0.03T^2 - 0.7F + 1034.5$ F: fat content (%) T: temperature (°K)

Human milk rheological studies are limited. An early work by Blair noted shear-thinning behavior but determined the decrease to be so minor that milk could be classified as Newtonian [19]. Blair's work was limited to shear rates above 100 s^{-1} but provided the basis for a major study on viscosity of raw human milk by Waller et al. Assuming that raw human milk was Newtonian, Waller et al. explored the kinematic viscosity of human milk during the first 10 days of lactation, when the composition of milk, particularly proteins, changes the most [20]. They tested the samples at 37 °C when milk fat was liquid [18]. Kinematic viscosity significantly decreased during the first 10 days postpartum which corresponded to the drop in total nitrogen content. A linear relationship between the log of kinematic viscosity, ν, and total nitrogen content, c, was determined and expressed as $\log \nu = 0.65c - 0.07$.

Waller et al. [20] further investigated the decrease in kinematic viscosity by exploring the relationship between casein and globulin, two known protein nitrogens, which undergoes a significant change during the first 14 days of lactation before becoming almost constant. This work demonstrated a relationship between human milk content and viscosity yet is difficult to compare with other works. Handling of samples prior to testing was inconsistent as some samples were fresh while others were previously refrigerated, and time between collection and testing varied. Also, density differences were not disclosed nor discussed.

A few more recent studies examined human milk viscosity with most research assuming Newtonian flow behavior [21–26]. Almeida et al. studied human milk viscosity in regards to aging as it pertains to clinical treatment of infants with dysphagia and determined no significant changes occurred when previously frozen human milk was reheated and then maintained at 37 °C over 9 h [21]. A subsequent study in regards to managing infant dysphagia tested previously frozen human milk samples from 2 donors at 25 °C from 1–1000 s^{-1} and found shear-thinning with greater variability in viscosity when compared with infant formula [23]. Another study concerning lipid digestion by infants noted shear-thinning using a logarithmic shear rate sweep from 0.5–500 s^{-1} at 37 °C on previously refrigerated human milk [22]. In a separate test conducted at an arbitrarily chosen 20 s^{-1}, they dropped milk pH from 6.5 to 4.0 with no changes in viscosity. The authors did not show the results of the viscosity testing nor was density mentioned.

In light of the importance of human milk to the health and development of infants, this work aims to explore the general rheological behavior of human milk in response to temperature, storage (refrigeration/freezing), aging (constant body temperature), time (thixotropy), and shear rate with additional consideration of the intra-individual milk content variations between breasts of the same woman (inter-breast) and over the course of a single expression of milk by pump or infant suckling (intra-feed). Particular attention is given to low shear rates and body temperature, as experienced in tube feeding [4,5,27,28].

2. Materials and Methods

2.1. Recruitment and Milk Collection

A total of 8 participants were recruited and provided informed consent at different points in the study (for further details concerning recruitment and milk collection procedures, please refer to [29]). The Institutional Review Board of University of Texas at Tyler approved the study (IRB 15–10). Participants #1–6 expressed simultaneously from both left (L) and right (R) breasts continuously until milk flow stopped using double electric vacuum pumps. Participants who expressed larger volumes had 20 mL separated from total volume for testing unless participant opted to donate the entire expression (Participants #2, #5, & #6). Participants #7–8 expressed approximately 15 mL at the beginning of their expression/nursing (hereafter referred to as foremilk (F)) and additional milk at the end of their expression/nursing (hereafter referred to as hindmilk (H)). Participant #8 expressed using a double electric vacuum pump while Participant #7 hand expressed before and after breastfeeding her infant only from the suckled breast. In total 15 raw human milk samples were obtained ranging in volume from 5 mL to 52 mL. Macronutrient content was calculated as outlined in Appendix B. All fresh samples were tested within 5 h of expression with most tests beginning within 30 min. Any volume not aliquoted for fresh testing was immediately refrigerated/frozen as a whole (not in aliquots). Refrigerated samples were held at 4 °C for up to 5 days while frozen samples were held at −20 °C for 3 months until thawed in a warm water bath between 30–40 °C. After thawing, samples were held at 4 °C for up to 5 days.

2.2. Experimental Methods

The procedures for density and viscosity testing are detailed by Alatalo and Hassanipour [29] with an overview provided below. Density (ρ) was measured at temperature equilibrium with an Anton-Paar DMA 4500M density meter with an accuracy rating of ±0.00005 g cm^{-3} and ±0.03 °C. To study the effect of storage, density was calculated at a single temperature for Participants #1–6 within 10 min after expression and compared with the results of density after freezing and thawing. The density of the foremilk and hindmilk of Participant #8 was tested the same day as expressed and after refrigeration at 4 °C to determine the effect of time of expression (hereafter referred to as intra-feed variations) and storage by refrigeration. Testing parameters for experiments on the density meter are outlined in Table 2.

Table 2. Density Testing Details.

Parameter	Test # 1	Test # 2
Temperature Range	3–50 °C	36–41 °C
Tested Sample(s)	#1–6 *	#8 **
Total Data Points	40	6
Measured At	Sample Equilibrium	
Recording Frequency	Every °C	
Sample Volume	1.3 mL	

* After freezing; ** Fresh.

Viscosity experiments were performed using an Anton-Paar MCR-302 rheometer. Shear-dependent rotational experiments used a cone-plate measuring system with Peltier temperature controlled plate and hood. The cone measured 49.9384 mm in diameter with a 50 µm truncation and 0.49° angle. Since the focus of this study is on low shear rates and sample availability was limited, the cone-plate measuring system allowed for more experiments as testing required loading volumes of 0.4 mL trimmed to 0.29 mL. The test parameters were initially determined by testing whole raw bovine milk purchased from Lavon Farms, Plano, Texas, USA, and are outlined in Table 3. Two separate shear rate ranges were tested. For both tests, samples were brought to 37 °C and held at that temperature for 4 min before applying shear. Body temperature was chosen due to the practice of warming infant feeds to body temperature, particularly in fragile and preterm infants [27,28]. Three separate temperature sweep ranges were tested using a linear ramp profile of 1 °C every minute. All tests occurred with a 4 min pre-shear of $50\ \mathrm{s}^{-1}$ at the initial temperature for the individual sweep. The viscosity was read after the pre-shear and compared with the first data reading of the sweep. This comparison was made to ensure that temperature was the only variable affecting the viscosity readings during the sweep. The shear rate of $50\ \mathrm{s}^{-1}$ was chosen based on the shear rate point used by the National Dysphagia Diet for classification of food thickness and to allow for comparison with results published by Frazier et al. [23]. The loop test followed the same 4 min rest period at 37 °C as the shear rate tests.

Oscillatory experiments on the Anton-Paar MCR-302 were performed using a double gap cylinder measuring system with Peltier temperature control. The use of a double-gap cylinder measuring system requires greater sample volume, 4 mL, but significantly increases the available shear area enabling uniform shear conditions on both the inner and outer walls and detects lower torques better compared to other measuring systems [30]. The bob effective length, inner diameter and outer diameter are 40, 24.66 and 26.66 mm respectively. The cup inner diameter is 23.826 mm and outer diameter is 27.592 mm. Samples from Participants #2, #5, and #6 from both left and right breasts were heated to 37 °C and held at that temperature for 4 min before applying shear strain. The oscillating shear strain was applied using a logarithmic ramp from 0.01% to 1000% with a constant angular frequency of $5\ \mathrm{rad\ s}^{-1}$ which approximates infant suckling [31]. The duration of the test was set by the device. Testing parameters for all experiments on the rheometer are summarized in Table 3.

The complex shear modulus, $G^\star$, was broken down into its two components: (1) G'—the storage modulus that characterizes the elastic behavior—and (2) G''—the loss modulus that characterizes the viscous behavior. The limit of the linear viscoelastic (LVE) region, γ_L, was used to determine the yield point, τ_y, for each sample. For consistency between samples, γ_L was determined at the highest G' value before the slope of the curve became negative. The flow point, τ_f, was determined from the crossover points for G' and G'' where viscous behavior begins to dominate. While some authors associate the crossover stress and strain as the yield point, this association is inaccurate since G' and G'' are only valid in the LVE region [32]. The final characteristic of raw human milk considered in this study was the flow transition index (FTI). The FTI is the ratio of τ_y to τ_f and describes the transition behavior of the milk from the LVE region until flow begins.

Table 3. Viscosity Testing Details.

Parameter	Shear Rate Sweep		Temperature Sweep			Shear Rate Loop	Amplitude Sweep
	Test #1	Test #2	Test #1	Test #2	Test #3		
Temperature(s)	37 °C	37 °C	0–50 °C	29–45 °C	36–43 °C	37 °C	37 °C
Sample(s)	#1–6 *, 8 **	#1–6 *, 8 **	#1–6 *, 8 ***	#1–6 *	#7 **	#8 ***	#2 *, 5 *, 6 *
# Data Points	100	40	51	33	8	400	25
Measured At	Linear Ramp 1–100 s^{-1}	Linear Ramp 0.01–20 s^{-1}	50 s^{-1}	50 s^{-1}	50 s^{-1}	1–200–1 s^{-1}	0.01–1000% at ω = 5 rad s^{-1}
Point Density	1 $\dot{\gamma}^{-1}$	2 $\dot{\gamma}^{-1}$	1 °C 60 s^{-1}	1 °C 30 s^{-1}	1 °C 60 s^{-1}	1 $\dot{\gamma}^{-1}$	6 decade^{-1}
Recording Frequency	Constant Every 2 s	Linear Ramp 10–1 s	Constant Every 60 s	Constant Every 30 s	Constant Every 60 s	Constant Every 1 s	Set By Device
Volume	0.29 mL	0.29 mL	0.29 mL	0.29 mL	0.29 mL	0.29 mL	4.0 mL
Test Time	440 s	460 s	55 min	21 min	12 min	640 s	Varied

* After freezing only; ** Fresh & After refrigeration; *** After refrigeration only.

2.3. Uncertainty Analysis

Limited sample volume often restricts the number of times experiments can be repeated. When dealing with a biofluid, the composition from an individual is in constant flux and would require either pooling samples or limiting experiments to single-samples (no repetition). One of the goals of this work is to explore the breadth of rheological behavior of human milk normally found in nature. In keeping with that purpose and the limited sample sizes noted in Section 2.1, repetition of experiments was minimized. Since density is assumed to be incompressible, the accuracy of the meter was considered sufficient for uncertainty. To estimate the uncertainty of rheological single-sample experiments, the Kline-McClintock method [33] was employed and described herein.

The Anton-Paar MCR-302 rheometer has torque resolution of 0.1 nNm, angle resolution (determined from displacement by optical encoder) of 10 nrad, temperature resolution of 0.1 °C, and time constant of 5 ms. Viscosity measurements for cone-plate system are calculated from McKennell [34]:

$$\mu = \frac{3\alpha T}{2\pi r^3 \omega} \tag{1}$$

where α is cone angle, T is measured torque, r is cone radius, and ω is rotational speed. The uncertainty of the cone geometry is $\pm 0.005°$ for angle and ± 0.00005 mm for diameter. Using Kline-McClintock [33] analysis, the uncertainty for the cone-plate measuring system are determined using:

$$w_\mu = \sqrt{\left(\frac{3 \times 10^{-10}\alpha}{2\pi r^3 \omega}\right)^2 + \left(\frac{10^{-3}T}{24 r^3 \omega}\right)^2 + \left(\frac{-1.5 \times 10^{-8}\alpha T}{\pi r^3 \omega^2}\right)^2 + \left(\frac{-7.5 \times 10^{-8}\alpha T}{\pi r^4 \omega}\right)^2} \tag{2}$$

with $r = 0.0249692$ m and $\alpha = 0.49°$ for all rotational tests. The vast majority of w_μ remained under 1.03% with less than 10 data points having a higher uncertainty at $\dot\gamma \leq 1\,\text{s}^{-1}$. The maximum uncertainty was 8.20% for $w_\mu = 0.3$ mPa s at $\dot\gamma = 0.01\,\text{s}^{-1}$.

3. Results

3.1. Milk Content

The macronutrient content of milk samples, presented in Table 4, was assumed to match reference values from clinical studies. These values do not account for the natural variations found between breasts of the same participant [35] but provide reasonable expected values for evaluation of data. Further discussion of these values is provided in Appendix B.

Table 4. Estimated Macronutrient Content.

Participant	Month of Lactation	Carbohydrates (g/100 mL)	Proteins (g/100 mL)	Fats (g/100 mL)
#1	9.0	7.01	0.80	4.11
#2	16.0	6.80	0.97	5.23
#3	12.5	6.91	0.90	4.67
#4	12.25	6.91	0.89	4.63
#5	3.5	7.18	0.66	3.23
#6	4.25	7.16	0.68	3.35
#7F	8.0	7.04	0.78	3.17
#7H	8.0	7.04	0.78	6.61
#8F	1.0	7.26	0.60	2.25
#8H	1.0	7.26	0.60	4.81

3.2. Flow Behavior of Human Milk

Human milk demonstrates shear-thinning non-Newtonian behavior in response to increasing shear rates (see Figure 1a) as reported regarding human milk by [22,23]. While the slope of the curve decreases at higher shear rates, viscosity never becomes constant over the range of shear rates tested. Conversely, as shear rate approaches 0 s^{-1}, viscosity approaches infinity. This flow behavior is especially evident in the Shear Rate Sweep Test #2 results in Figure 1b, which began at 0.01 s^{-1} and provides greater detail of rheological behavior at or near zero. The standard deviation (SD) for milk viscosity was largest at lower shear rates (± 683.6 mPa s at 0.01 s^{-1}) and consistently decreased as shear rate increased (± 0.7 mPa s at 100 s^{-1}). By 61 s^{-1}, the standard deviation was ≤ 1 mPa s. While milk with the lowest viscosity values fell within the expected range of standard deviation, the highest viscosity milk remained outside the standard deviation range for the entire sweep. Some milk demonstrated an oscillatory viscosity pattern at low shear rates, generally between 10–60 s^{-1}, that was obscured when calculating the mean viscosity for the samples [29]. Two likely sources for these viscosity patterns are deformation of fat globules, which usually range in size from 1–10 µm, and the breakdown of casein micelles (comprising approximately 13% of total protein) into smaller micelles [36] similar to breaking up red cell aggregation in blood [37].

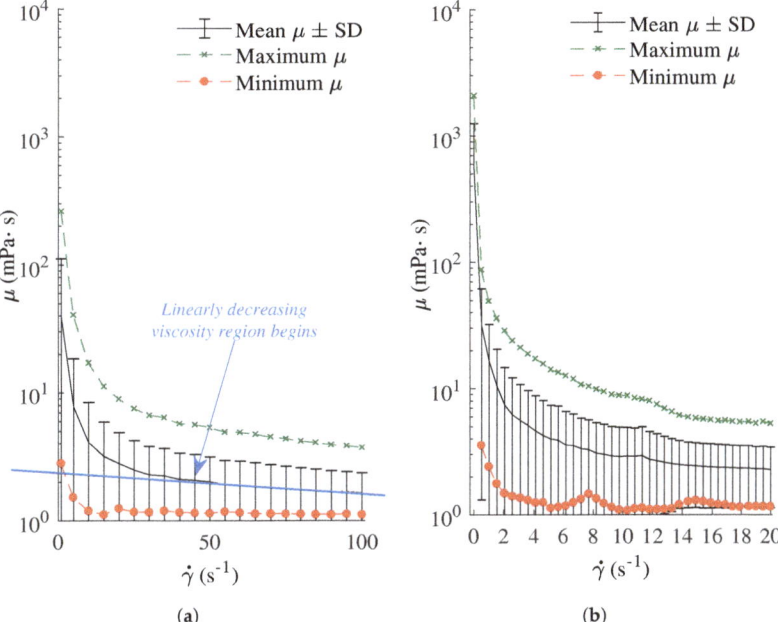

Figure 1. Shear-thinning flow behavior of human milk. Raw human milk demonstrates shear-thinning flow behavior with μ decreasing as $\dot{\gamma}$ increases. The maximum and minimum μ values for each Shear Rate Sweep Test are presented to show the range of data in comparison with the mean and SD. (**a**) Shear Rate Sweep Test #1 (1–100 s^{-1}) shows a continuously decreasing slope that never reaches Newtonian. (**b**) Shear Rate Sweep Test #2 highlights $\lim_{\dot{\gamma} \to 0} \mu \to \infty$.

To determine the presence of time dependence for the viscosity, the Shear Rate Loop Test from 1 s^{-1} to 200 s^{-1} and back to 1 s^{-1} was performed on the 5-day post-refrigerated milk from Participant #8. The result of each sample was similar with a maximum uncertainty of 0.22 mPa s calculated from Equation (2) (see Figure 2) and confirmed human milk to be a thixotropic fluid.

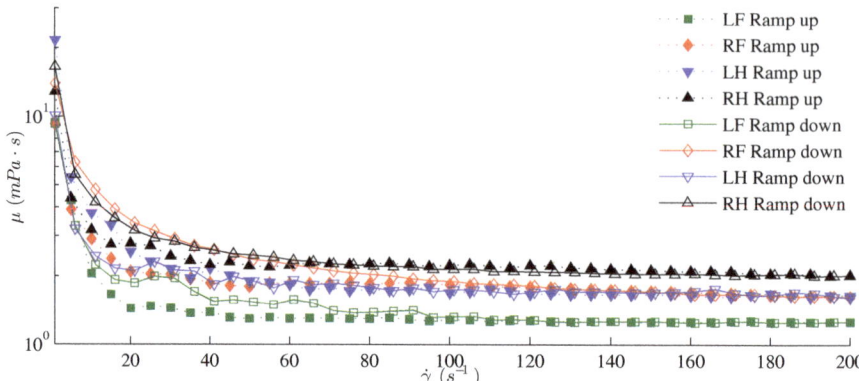

Figure 2. Evidence of time dependence for human milk flow properties. Shear Rate Loop Test demonstrates time dependence for human milk for both foremilk and hindmilk. Viscosity of milk from the right breast ended higher while milk from the left breast ended lower and shows inter-breast variation (see Section 3.4).

Rheological rotational tests where viscosity approaches infinity as shear rate approaches 0 s^{-1} indicate the existence of a yield stress that defines a plastic fluid. This behavior was seen in every Shear Rate Sweep Test for human milk, which suggests that raw human milk has a yield point (i.e., plastic fluid) and firm texture when at rest. The Amplitude Sweep Test with controlled shear strain allowed for calculation of shear stress, phase shift, complex modulus, storage modulus, and loss modulus that confirmed the viscoelastic behavior of raw human milk, particularly as it pertains to infant suckling behavior. All samples had G' values greater than G'' in the LVE region as shown in Figure 3a for Participant #6, although the differences between G' and G'' were small (0.2936 ± 0.5200 Pa). This relationship classifies raw human milk as a viscoelastic solid (gel-like). Since G' increased linearly for some samples in the LVE region, such as Sample #6L in Figure 3a, the yield point was confirmed using the stress-strain curve as shown in Figure 4. The increase in G' within the LVE region indicates a strengthening of structure under small amplitudes. Increases in intermolecular crosslinking, aggregation, and particle size of proteins within raw milk has been found in response to increases in shear rate below 500 s^{-1} which likely accounts for the increase in G' [38]. The mean γ_L was $1.10 \pm 0.76\%$.

The shear stress necessary to begin breakdown of structure and initiate flow, τ_y, varied between samples with a mean value of 13.903 ± 26.901 mPa showing a wide range of expected values. Inter-breast differences ranged from a factor of 5 to a factor of 100 with each participant having milk from one breast that required <1 mPa to initiate yielding (intra-individual variations discussed further in Section 3.4). Homogeneous flow begins at τ_f when G' crosses G'', as shown in Figure 3a for Participant #6, allowing for viscous dominated flow behavior. The mean τ_f was 24.28 ± 35.68 mPa. Both τ_y and τ_f can be seen in Figure 3b. The mean difference between τ_y and τ_f for each sample was 10.38 ± 10.01 mPa. FTI values, which describe the transition behavior of milk from the LVE region τ_f, ranged from 1.39 to 22.17 with mean FTI of 6.34 ± 8.08. A summary of these points are provided in Table 5.

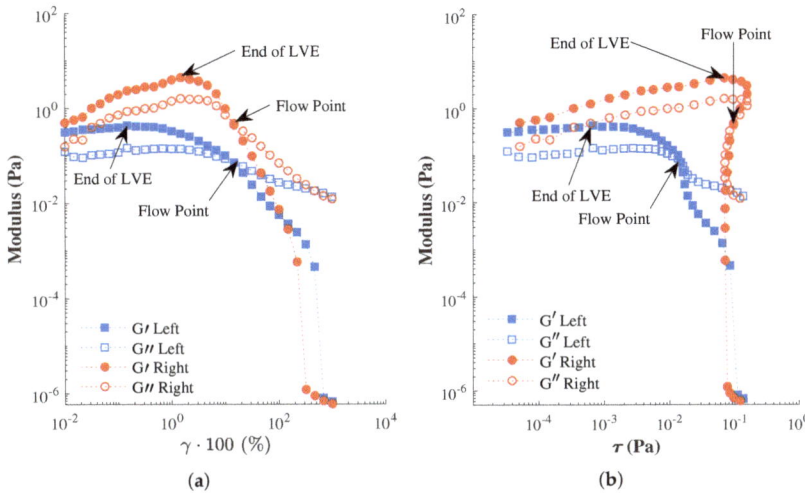

Figure 3. The viscoelastic behavior of human milk represented by milk from Participant #6. (**a**) The deformation at the limit of LVE, γ_y, and crossover of G' and G''. (**b**) The corresponding shear stress required for yield and flow of human milk.

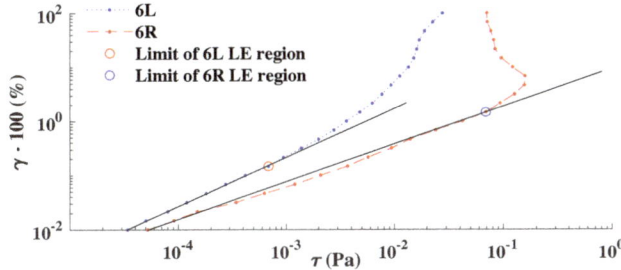

Figure 4. Determining the yield point by finding the limit of the linear-elastic (LE) region. When a straight line is fitted to the linear region of the stress-strain curve, where the Law of Elasticity applies, the final point before the yield exceeds the line corresponds to the end of the LVE shown in Figure 3.

Table 5. Limit of LVE region (γ_L), yield point (τ_y), flow point (τ_f), and flow transition index (FTI) values for six samples.

Sample	Density (g cm^{-3})	γ_L (%)	τ_y (mPa)	τ_f (mPa)	FTI
2R	1.02432	1.47	8.428	22.37	2.65
2L	1.02656	0.32	0.220	1.68	7.63
5R	1.03002	1.00	0.891	2.00	2.24
5L	1.02604	2.16	4.766	9.26	1.94
6R	1.02210	1.47	68.433	95.30	1.39
6L	1.02795	0.15	0.680	15.07	22.17
mean ± SD		1.10 ± 0.76	13.903 ± 26.901	24.28 ± 35.68	6.34 ± 8.08

3.3. Effect of Temperature on Human Milk Density and Viscosity

The effect of temperature on density for the first 12 samples post freezing, expressed as specific gravity (ρ_{milk}/ρ_{water}) in Figure 5, shows a decrease as the temperature increases. The density decrease was approximately 0.02 g cm^{-3} over 47 °C for each sample, although the decrease is not linear. The slopes for specific gravity curves decrease above 20 °C and indicate that at lower temperatures volume changes of components within the milk exert a larger influence on density than at higher temperatures.

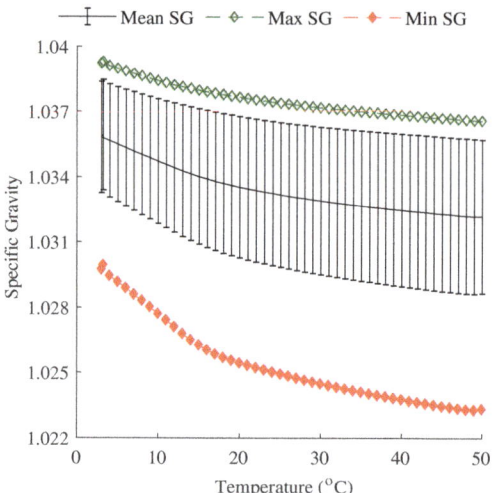

Figure 5. Human milk density response to temperature. Specific gravity (SG) (ρ_{milk}/ρ_{water}) shows greater changes in lower temperature range. The maximum and minimum SG values both fall outside the expected standard deviation (SD) range shown in the figure.

Temperature influences on viscosity were detected by Temperature Sweep Tests #1 and #2. The pre-shear viscosity value was compared to the first viscosity recorded during the actual temperature sweep with only 2 pre-shear viscosity values varying greater than 1 mPa s compared to the first data point in the sweep. These results indicated that 4 min was sufficient time for the milk particles to orient themselves and to isolate the effect of temperature on the viscosity. Temperatures greater than 50 °C were avoided to prevent denaturization of proteins that would alter viscosity [14]. In general an increase in temperature decreased the viscosity. However as shown in Figure 6a, some samples demonstrated an increase in viscosity between 6–16 °C before returning to anticipated behavior resulting in the mean temperature rise seen between 8–11 °C. This range corresponds to the slope change in specific gravity seen in Figure 5. All the samples that were tested to 50 °C demonstrated an increase in viscosity beginning as early as 41 °C for some samples and all samples by 46 °C. Temperature Sweep Test #2 used a new loading of sample from Participants #1–6 over a smaller temperature range (29–45 °C) and faster recording frequency to shorten the total testing time and ensure that viscosity increases above 40 °C were not related to sample drying due to the long test time in Temperature Sweep Test #1. The results from both temperature sweeps were averaged over normal milk reheating temperature ranges and presented in Figure 6b.

A linear approximation of each individual Temperature Sweep Test #1 and #2 result for Samples #1–6 was completed in MATLAB using the "fit" function and polynomial model "poly1". The mean ± SD slope from 36–40 °C was −0.11941 ± 0.13916. All but one sample had a negative slope indicating a decrease in viscosity as temperature increased. The slope for Temperature Sweep Test #2 (29–45 °C) result for Sample #6L was positive (0.06988) which prompted a second test with a new loading that also resulted in a positive slope (0.1078). Both the density and specific gravity slopes over that same temperature range were unremarkable for Sample #6L. The slope for Temperature Sweep Test #1 (0–50 °C) for Sample #6L was negative (−0.07364) as expected. The only noted differences between the tests were in point density and recording frequency. The possibility exists that Temperature Sweep Test #2 for Sample #6L with repetition contained experimental errors since the slopes were inconsistent with all other results, although the positive slopes were so small that the resultant viscosity values, which increase with temperature increase, fail to be significant.

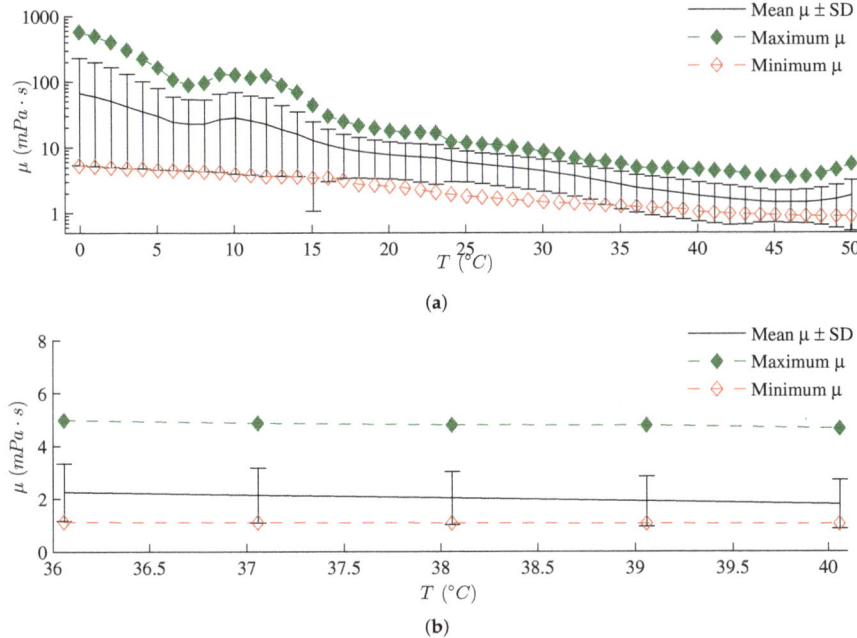

Figure 6. Human milk viscosity response to temperature at 50 s^{-1}. Human milk viscosity generally decreased as temperature increased. The highest (maximum) and lowest (minimum) viscosity values are presented to show the range of recorded data with maximum viscosity falling outside the expected value range of SD. (**a**) Unexpected increases in mean and maximum viscosity began at 8 °C and 46 °C. (**b**) While temperature decreased viscosity over normal reheating temperature ranges, the decrease was minimal.

3.4. Changes in Density and Viscosity Associated with Intra-Individual Human Milk Variations

Inter-breast density varied among participants and likely stems from the known inter-breast variations in milk composition, which can be significant [35] (further discussion in Appendix B). At 37 °C the difference between breasts ranged from as high as 0.00730 g cm^{-3} for Participant #1 to as low as 0.00011 g cm^{-3} for Participant #4. The mean difference was 0.00289 g cm^{-3} with details for each participant shown in Table 6.

Table 6. Human milk, after thawing from the same participant but different breasts, shows differences in ρ (g cm^{-3}) at 37 °C.

Participant	Right Breast	Left Breast	Difference
1	1.02443	1.01713	0.00730
2	1.02432	1.02656	0.00224
3	1.02583	1.02883	0.00300
4	1.02750	1.02761	0.00011
5	1.03002	1.02604	0.00398
6	1.02210	1.02795	0.00585
Mean ± SD			0.00289 ± 0.00269

The mean±SD density of fresh milk foremilk and hindmilk for Participant #8 is shown Figure 7 and reflects intra-feed variations that correspond to intra-feed variations in fat seen in Table 4. Milk reheating practices and temperatures can vary [28], so the small temperature range around 37 °C in Figure 7 highlights the degree of change in density with small fluctuations in temperature. All fresh samples were held at 37 °C until time of testing. The difference between fresh foremilk and hindmilk density in Figure 7 averaged 0.00489 g cm^{-3} in the right breast and 0.00464 g cm^{-3} in the left over the temperature range tested.

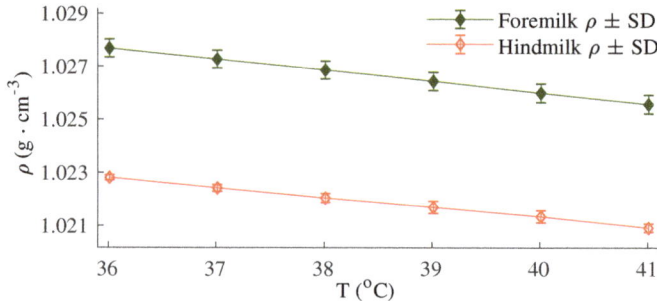

Figure 7. Density Test #2 for samples from Participant #8 for intra-feed (foremilk/hindmilk) and inter-breast (shown with SD caps) variations tested fresh. While higher fat content lowers density, inter-breast differences are minimal, particularly for hindmilk.

Temperature Sweep Tests #1 and #3 for Samples #8 and #7, respectively, show viscosity differences (Figures 8 and 9) correspond to intra-feed variations in fat content with hindmilk consistently having higher viscosity compared to foremilk from the same breast and feed. Hindmilk viscosity increased when the temperature increased beginning at 41 °C for Sample #7H (Figure 8) and at 44 °C and 47 °C for Samples #8LH and #8RH, respectively, (Figure 9a). Foremilk failed to show any increase in viscosity suggesting that the increase is associated with higher fat content, particularly since the highest fat content milk, Sample #7H, also showed the earliest increase in viscosity at 39.5 °C. This finding merits further investigation considering possible implications for infant feeding. A closer look at viscosity over reheating temperature range showed hindmilk was higher than foremilk (see Figure 9b) with the Sample #8R hindmilk twice the viscosity value of either Sample #8 foremilk sample.

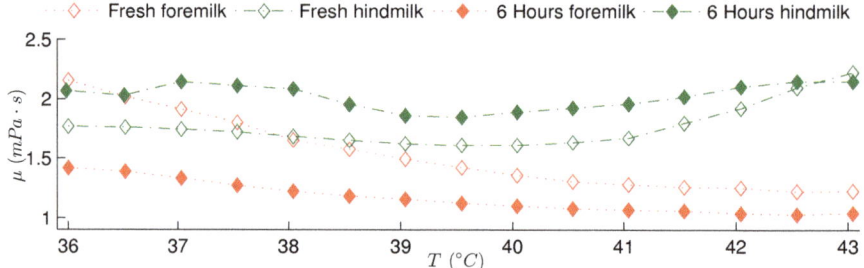

Figure 8. Temperature Sweep Test #3 at $\dot{\gamma}$ = 50 s^{-1} shows intra-feed variations in viscosity that relate to composition. Hindmilk consistently has higher viscosity compared to foremilk from the same breast and expression (intra-feed variations). When the samples were aged at 37 °C for 6 h (filled diamonds), the foremilk viscosity decreased slightly while hindmilk increased (see Section 3.5). Samples from Participant #7.

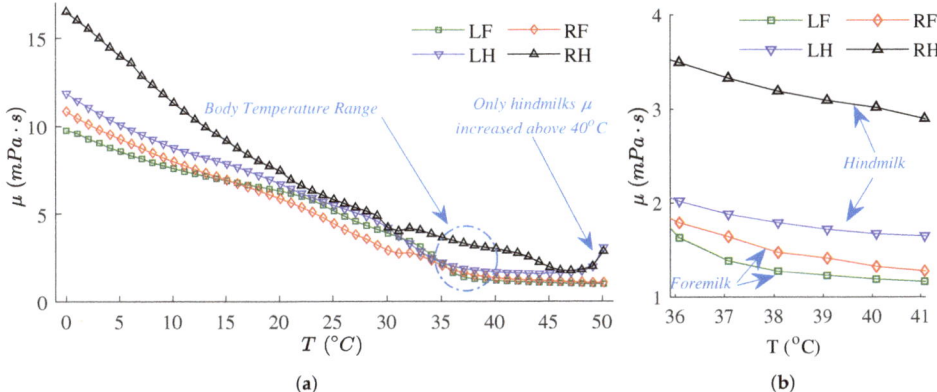

Figure 9. Temperature Sweep Test #1 for samples from Participant #8 at $\dot{\gamma}$ = 50 s^{-1} after 5 days refrigeration. (**a**) Temperature decreased viscosity for both foremilk and hindmilk until 45 °C when hindmilk began to increase, (**b**) A closer look over body temperature range shows a steady decrease in viscosity in response to temperature.

Shear Rate Sweep Tests #1 and #2 for samples from Participant #8 showed intra-feed variations particularly at low shear rates <10 s^{-1}. At $\dot{\gamma}$ = 0.01 s^{-1}, fresh foremilk viscosity exceeded fresh hindmilk viscosity by more than one order of magnitude, which can be seen in Figure 10a. At that same shear rate the inter-breast viscosity variations for foremilk (right breast dominated) exceeded the differences for hindmilk (left breast dominated). As shear rate approached 100 s^{-1}, the intra-feed variations did not remain consistent. As seen in Figure 10b, fresh foremilk from the right breast dominated all other samples yet viscosity for fresh foremilk from the left breast was lower than other samples, including previously refrigerated ones (for further discussion regarding storage effects, see Section 3.5). These results clearly show the influence of proteins at low shear rates on milk rheology similar to research on raw skim bovine milk [6].

Inter-breast and intra-feed thixotropic viscosity variations appeared in Shear Rate Loop Test. A comparison of start and end viscosities at 1 s^{-1} for the loop test shows that milk from the left breast decreased viscosity while milk from the right breast increased viscosity. The percentage of increase or decrease is presented in Table 7. The differences in fat content between foremilk and hindmilk do not appear to be a factor in determining whether viscosity increases or decreases with time.

Table 7. Start and end viscosities at 1 s^{-1} for the loop test.

Sample	Initial μ (mPa s)	Final μ (mPa s)	Increase/Decrease (%)
Left Foremilk	9.4136	9.2705	−1.52
Right Foremilk	9.3010	13.9690	+50.19
Left Hindmilk	21.6380	10.1380	−53.15
Right Hindmilk	12.8800	16.6060	+28.93

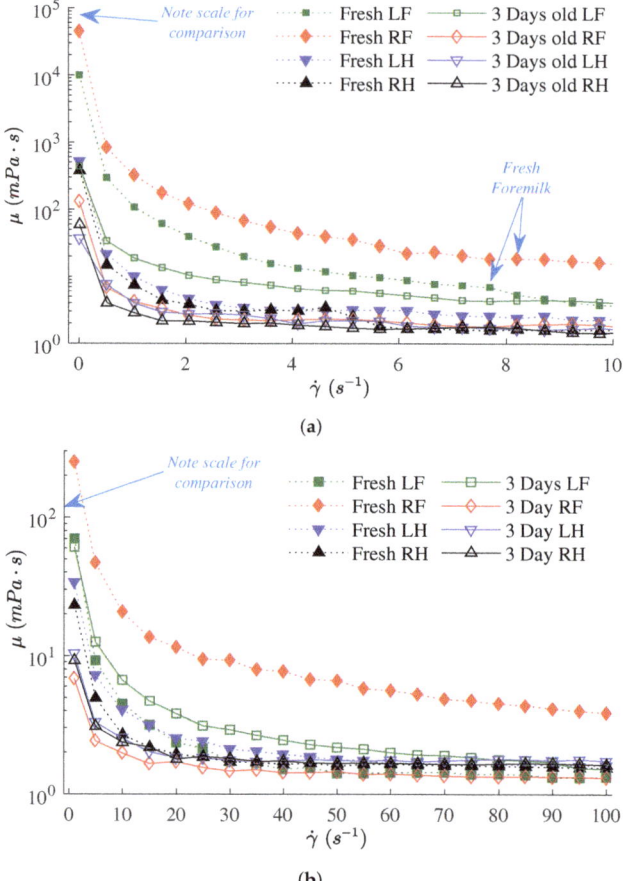

Figure 10. Shear Rate Sweep Tests #2 (Figure 10a) and #1 (Figure 10b) results for milk from Participant #8 show both inter-breast and intra-feed variations in viscosity. Effect of 3 days of refrigeration on milk viscosity associated with intra-individual variations is also seen (discussed in Section 3.5). (**a**) The low shear rates experienced during the beginning and end of a suck cycle show higher viscosity in fresh foremilk (lower fat) compared to hindmilk (higher fat). Storage by refrigeration lowered viscosity of each sample until crossover for foremilk at $8\,s^{-1}$ and hindmilk at $6\,s^{-1}$. Inter-breast variations were greater for fresh foremilk than fresh hindmilk and for fresh milk compared to stored milk. (**b**) The starting viscosity (at $\dot{\gamma} = 1\,s^{-1}$) for all fresh samples was higher than after 3 days of refrigeration but that pattern did not continue as shear rate increased.

3.5. The Effect of Storage & Aging on Human Milk Density and Viscosity

Density measurements for both fresh and previously frozen milk are shown in Table 8. The average density for fresh milk, after removing the outlier Samples 3R and 3L, was $0.964\,g\,cm^{-3}$ with a standard deviation of $0.007\,g\,cm^{-3}$. The source of deviation for fresh Samples 3R and 3L is assumed to be due to experimental error since they were tested at the same time. A 95% confidence interval for both fresh and thawed density mean was calculated using a t distribution. The results show an increasing in density for thawed milk (average 6.8%) when compared with fresh milk measurements at the same temperature, with the samples having the lowest fresh density increasing the most.

Table 8. Human milk density increased in response to storage at $-20\,^\circ\text{C}$ for 3 months. Fresh samples were tested within 20 min of expression in a single reading at the milk temperature at time of arrival to equipment. The thawed values for comparison were recorded in Density Test #1. The fresh results for milk from Participant #3 were outliers due to assumed experimental error and removed from these test results.

Sample	Temperature (°C)	Fresh ρ (g cm^{-3})	Thawed ρ (g cm^{-3})
1R	26.4	0.970	1.029
1L	24.8	0.964	1.022
2R	25.8	0.967	1.028
2L	26.7	0.964	1.030
4R	26.0	0.972	1.031
4L	27.0	0.969	1.031
5R	24.8	0.952	1.034
5L	25.4	0.962	1.030
6R	26.2	0.968	1.026
6L	26.0	0.952	1.032
Mean ± SD	25.9 ± 0.74	0.964 ± 0.007	1.029 ± 0.003
95% Confidence Interval		$0.959 \leq \rho_{mean} \leq 0.969$	$1.008 \leq \rho_{mean} \leq 1.050$

Patterns in viscosity changes due to storage by refrigeration and aging were determined with milk from Participants #7 and #8. Participant #7's milk was tested at a constant shear rate of 50 s^{-1} over a small temperature range representative of milk reheating temperatures both fresh and after aging at a constant 37 °C for 6 h similar to what [21] performed. Aging effects can be seen in Figure 8. Hindmilk viscosity increased with aging while foremilk viscosity decreased.

Participant #8's milk samples were held at 37 °C and not initially tested until 5 h after collection. Due to this delay, data regarding changes due to aging at 37 °C were not obtained. All samples from Participant #8 were tested at 5 h after expression and after 3 days refrigeration at 4 °C using Shear Rate Sweep Tests #1 and #2. For the fresh and 3 days post-refrigeration tests, each foremilk and hindmilk sample was tested twice with a new loading and then averaged. The results in Figure 10a show that the fresh milk samples tested considerably higher at very low shear rates (Shear Rate Sweep Test #2) compared to post-refrigeration samples. This pattern does not continue for all samples as shear rate increases. Figure 10b shows that at 1 s^{-1} milk viscosity was greater when fresh than post-refrigeration, yet by 100 s^{-1} post-refrigeration samples were greater for all but right breast foremilk, which maintained a significant difference between fresh and post-refrigeration results throughout testing.

4. Discussion

To date there is very little rheological data pertaining to raw human milk, which is not surprising considering early works determined human milk to be approximately Newtonian [19] and few engineering models of breastfeeding have been produced [39–41]. In this study, freshly expressed human milk was tested to determine general behavior patterns and alterations in density and viscosity in response to environmental factors, storage, aging, and intra-individual milk content variations.

Density is frequently used by clinicians and in research studies when estimating milk intake (mass or volume) for infants by assuming density is close to 1 g cm^{-3} [42–44]. Since human milk is a dynamic biofluid with over 100 components that show intra-feed variations during each breastfeed [35,36,45–51] and factors such as temperature and storage affect density, understanding the range of variation can help ensure more accurate estimations of milk intake. The results of this study show that in the clinical setting where test-weighing is used to estimate infant intake, this approximation is sufficient considering infant scale accuracy is often ±1 g or more. Inter-breast differences in milk density existed and likely originated from composition differences [35,46]. The influence of intra-feed composition changes on density are evident when comparing foremilk and hindmilk. Since fat concentration varies the most over the course of a single expression/feed while lactose and protein remain fairly constant [47,52], the differences in density between foremilk and hindmilk are primarily due to fat fluctuations with higher fat content decreasing density.

Human milk flow behavior demonstrated consistent patterns of variation with regards to shear rate, temperature, storage, and aging when variations in macronutrient content was considered, although its thixotropic behavior, similar to blood [53], provided large range of expected values for some experiments. The Shear Rate Loop Test results were inconsistent in pattern with inter-breast variations since milk viscosity from one breast increased and from the other breast decreased as shear ramped down. The test was performed on post-refrigerated milk, so the effect of dissolved gases, that may be present in fresh milk, were minimized. However, chemical reactions and composition changes during storage (i.e., bacterial growth) could affect viscosity and be reflected in tests for thixotropy. In bovine milk studies, thixotropy was noted only in whole not skim milk [6] which suggests that the human milk fat globule has an associated relaxation time after deformation under shear. Since thixotropic behavior remained after storage, the fat globules maintained elasticity, although whether elasticity alters during storage is unknown but likely because lipolysis of human milk lipids occurs when milk is stored at any temperature above $-70\ °C$ [54].

The viscosity curve as a function of shear rate clearly shows non-Newtonian, shear-thinning flow behavior. As shear rate approaches $0\ s^{-1}$, viscosity approaches infinity indicating that raw human milk has a yield point and firm texture at rest, which was confirmed with oscillatory tests. The viscosity curve of individual samples showed an oscillatory behavior beginning around $20\ s^{-1}$ (see [29]) that smoothed as shear rate increased. The oscillations are likely due to changes in orientation, shape, and size of various milk particles. The decrease in slope at higher shear rates likely accounts for the many researchers who assumed Newtonian behavior for milk. However, currently there is insufficient research on human milk to define at what shear rate Newtonian behavior begins. Since milk is often infused at slow rates in narrow tubes, the non-Newtonian flow behavior is significant especially at low shear rates.

The non-Newtonian flow behavior at low shear rates also pertains to the oscillatory nature of suckling that constantly varies the pressure profile exerted upon milk when fed from artificial nipples. From the oscillatory testing, inter-breast variations in yield and flow points were determined. The disparities in yield and flow points indicate large differences in the structural strengths or gel strengths of the milk between breasts. This gel-like structure likely helps keep particles from settling when the milk is at rest. The decrease of the G' curves after the LVE region begins the nonlinear viscoelastic regime [32] and indicates a slow decline in structural strength, which demonstrates a smoother, more homogeneous flow behavior. A gradual rise in G'' after the LVE region occurred in 4 of the 6 samples tested and indicates that deformation energy was transformed into friction heat due to internal viscous friction while elastic behavior dominated. The high standard deviation shows the range of shear stress required to complete the necessary structural breakdown to initiate viscous dominated behavior.

Samples with FTI closer to 1 had a greater tendency for brittle fracturing or non-homogeneous breakdown to occur past the yield point. Sample 6R had the FTI closest to 1 and, as seen in Figure 3a, showed the steepest curve upon leaving the LVE region and had the highest G' value. So 6R experienced a large amount of internal viscous friction as it transitioned from elastic dominated to viscous dominated flow behavior. Other samples had FTI close to 1 with steep strain curves upon leaving the LVE region, so non-homogeneous breakdown appears common and may be correlated to fat content. Fat content is the only component with large intra-feed variations where pressure forces at the breast vary from multiple sources, like alveolar contractions and suckling. In tube feeding, fat is the component with the largest losses with significant protein losses also reported [4,5]. The loss of nutrients during tube feeding is likely due to non-homogeneous breakdown whereby a shear stress between the yield point and flow point is exerted causing a lower fat milk to flow. The apparent wall shear stress, ($\dot{\gamma}_a$), for tube feedings can vary from low (2.6 s^{-1}) to high (1697.7 s^{-1}) depending on flow rate and tube diameter (see Appendix C). Even though certain feeding parameters result in a high $\dot{\gamma}_a$, fat losses of over 50% are reported [5], which indicate pressure within the tube is insufficient to achieve homogeneous flow. Additionally, the thixotropic behavior of human milk adds an additional factor that likely contributes to feeding problems and requires further study.

The current recommendation for feeding preterm infants human milk is to warm milk to body temperature with the ideal temperature range of 35.5–37.2 °C although reported prefeed temperatures vary greatly (21.8–46.4 °C) [55]. Of particular interest in this study was the temperature range between 36–40 °C that immediately surrounds the normal body temperature. The mean density change between 36–40 °C was −0.0017 g cm^{-3} and viscosity decreased with fresh foremilk decreasing more compared to fresh hindmilk. However, higher temperatures showed areas of unusual flow behavior that could impact infant feeding. As noted in Section 3, viscosity began increasing for some samples as early as 41 °C and in all samples by 46 °C. The higher fat, hindmilk samples' viscosities began increasing at lower temperatures than their foremilk counterparts. Since feeding hindmilk to preterm infants is common in hospitals [48], small variations in heat have less impact on improving flow in tubing, but temperatures above 40 °C should be avoided.

The impact of storage on density and viscosity also suggested interesting changes in milk content. Past research tested milk after storage [18], but for the fresh samples in Table 8, density increased due to storage. Density for pure water is 0.9970–0.9965 g cm^{-3} for 25–27°, so freshly expressed milk density appears to be lower than water. One reason could be due to chemical changes [7,9] that altered the attraction between molecules and increased volume. Alternatively, milk may contain dissolved gases. The source of gases in human milk could be in vivo, being diffused from blood, or introduced during expression by vacuum pump. If the source is in vivo, then human milk is compressible, which theory requires further testing. Regardless of the source, with storage dissolved gases would be released leading to an increase in density. Dissolved gases would also impact viscosity behavior in fresh milk at low shears causing a higher resistance to flow and possible slip effects. This study found that refrigeration appears to decrease viscosity compared to fresh, particularly at low shear rates, which is supported by previous findings on bovine milk [6].

Lastly, the intra-feed variations associated with foremilk and hindmilk provided insight into the influence of fat content on milk viscosity. The hindmilk samples with higher fat content had lower viscosities than foremilk samples at low shear rates as shown in Figure 10a. This trend was also seen when the samples were tested over a higher shear sweep in Figure 10b. However, with the exception of fresh right-side foremilk, no extreme differences were seen at 100 s^{-1}. While the number of samples is insufficient to make absolute statements, the results demonstrate previously unreported flow behaviors. Since intra-feed protein concentration is stable, foremilk contains a higher protein to fat ratio. The protein composition during early lactation was found to heavily influence viscosity [20]. These results indicate that the protein content may have a stronger influence on viscosity at lower shear rates than previously thought and is supported by [38]. While testing for the milk yield point was not conducted on these samples, the extremely high viscosity values at 0.01 s^{-1} for the foremilk likely denotes higher internal cohesion forces when the protein to fat ratio is higher by allowing for formation of larger casein micelles that require greater shear stress to breakdown into smaller micelles to achieve flow.

5. Conclusions

Raw human milk flow properties vary with respect to temperature, storage, aging, time, and shear rate. Density of fresh milk when newly expressed may be lower than water due to dissolved gases that are released into the atmosphere during storage. A clear non-Newtonian flow behavior was found with large variations in viscosity, especially at low shear rates. The structural strength of milk varied with many samples showing non-homogeneous breakdown during the transition from yielding to flowing. This flow behavior can explain why tube feeding results in nutrient losses. The results highlight the need for sufficient pressure application to achieve homogeneous flow.

Author Contributions: Conceptualization, D.A. and F.H.; Methodology, D.A.; Software, D.A.; Validation, D.A.; Formal Analysis, D.A.; Investigation, D.A.; Resources, D.A. and F.H.; Data Curation, D.A.; Writing—Original Draft Preparation, D.A.; Writing—Review and Editing, D.A. and F.H.; Visualization, D.A.; Supervision, F.H.; Project Administration, D.A. and F.H.; Funding Acquisition, D.A. and F.H. All authors have read and agreed to the published version of the manuscript.

Funding: This material is based upon work supported by the National Science Foundation under Grant No.1454334 and 1707063, National Science Foundation Graduate Research Fellowship Program under Grant No.1746053, and Eugene McDermott Graduate Fellowship No.201701.

Acknowledgments: The authors thank Jimi Francis, University of Texas at Tyler, for her assistance with recruitment and collection, Maxine Quitaro and Anton-Paar for use of equipment, and all the mothers who donated to this study. Some preliminary findings were presented in 2016 ASME IMECE Young Engineer Paper Contest sponsored by the Fluids Engineering Division [29].

Conflicts of Interest: The authors declare no conflict of interest. The funders had no role in the design of the study; in the collection, analyses, or interpretation of data; in the writing of the manuscript; or in the decision to publish the results.

Appendix A

As the dairy industry grew, the need to model the relationship between viscosity and factors such as content and temperature led to the development of a series of regression equations. The usefulness of each equation depends on the application for which it was developed. Snoeren et al. found a relationship between the voluminosity of bovine milk proteins and the dynamic viscosity of commercial skim milk [10]. They showed that viscosity of heat-treated skim milk is a function of the volume fractions of casein, native whey protein, denatured whey protein, and the viscosity of the medium. Based on Bateman and Sharp's research on raw bovine milk [6], Snoeren et al. should have considered the shear rates used in testing due to the presence of native whey protein, which is found in raw bovine milk, yet no mention of shear rate is provided. Additionally, no data concerning the temperature of the milk used by Snoeren et al. was provided, even though studies show that temperature affects the viscosity of skim milk [14].

The effect of temperature and fat content on milk viscosity was investigated by Jebson and Chen [11]. They noted that whole milk has a higher solid content and higher viscosity in comparison to skim milk. In their research on the evaporation of bovine whole milk, for concentrates of solids content (larger that 450 g/kg), they adapted a relation for viscosity as a function of temperature and concentrate total solid. Similarly, Phipps [13] examined the relationship between viscosity, bovine cream fat content up to 50%, and temperature variations of 40 °C to 80 °C. He determined a regression equation for viscosity for creams with fat content (F) of less than 40%, as a function of temperature and fat content. Neither research considered shear rate in their experiments or equations.

More recent work with raw bovine milk was completed by Bakshi and Smith [12]. They performed several experiments on bovine milk to find a regression equation for the experimental value of viscosity based on the variable parameters of fat content and temperature. They noted that homogenized milk has a higher viscosity than raw milk which they attributed to the fine, dispersed state of the fat when homogenized. The temperature range tested was 0 °C to 30 °C and fat content range was 0.1% to 30%. Bakshi and Smith found the viscosities of skim milk and whole milk at 30 °C, to be about the same, approximately 1.25 mPa s. They also found that at lower temperatures, the effect of fat percentage on viscosity is greater. Similarly, they found a relationship between density, temperature, and fat content. All experimental work was performed at temperatures when milk fats are solid [18]. No regression equations were determined for raw milk nor was shear rate disclosed.

Appendix B

Human milk content varies in response to multiple factors. However, prior clinical studies provide some general guidelines that allow for estimation of content, which were used in this study.

Determining macronutrient content. Equations (A1)–(A3) determined by [49] based on month of lactation (n) to approximate macronutrient content (g/dL) are applicable under the following condition: the breast is fully emptied between the hours of 08:00 and 14:00. Participants #1–6 all met this requirement, so Equations (A1)–(A3) were used to calculate macronutrient content reported in the main text with limitations described later in this Appendix. For Participants #7–8, Equations (A1)–(A3) were used to calculate macronutrient content of their entire breast content and then modified to account for the normal variations found in foremilk and hindmilk based on percentages provided by [48]. The procedure for these modifications is outlined herein.

$$Carbohydrate = 7.2915 - 0.0309\,n \quad (A1)$$

$$True\ Protein = 0.5732 + 0.0258\,n \quad (A2)$$

$$Fat = 2.673 + 0.1597\,n \quad (A3)$$

Adjusting macronutrient content in foremilk and hindmilk. Over the course of a single feed, researchers [35,50] found that, between foremilk and hindmilk, protein levels remain the same, glucose decreases, and fat significantly increases 2–3 fold. Participants #7–8 both expressed milk after 14:00, however, since protein and carbohydrate content remains fairly constant throughout the day, only fat content was increased to account for the time of day [51]. Based on the aforementioned works, the assumption was made to exclude the changes in glucose since the molecular size is negligible when compared to fat globules, and macronutrient values were only adjusted for the fat content using the ratios found in [48] where foremilk has 82% and hindmilk has 161% of fat compared to composite or whole expression milk. Thus Equations (A1) and (A2) were employed to calculate carbohydrates and proteins, respectively. For fat, a composite value was found using Equation (A3), converted to g/L, and then used to calculate creamatocrit (%) using Equation (A4) [56]. The resulting creamatocrit value was then adjusted to foremilk or hindmilk based on [48] and converted back to g/dL.

$$Creamatocrit = 0.146 \times Fat + 0.59 \quad (A4)$$

Limitations of Equations (A1)–(A3). These equations do not take into consideration the natural variations between mothers nor the normal content differences between the breasts of the same mother, particularly fat. The differences between breasts for macronutrients involve fats and carbohydrates, primarily glucose. Work by [35] found significant fat variations in 60% of participants with results equally split between which breast expressed higher fat milk. The percent difference between breasts for those participants ranged from 25–72% with a mean difference of 46%. The significant fat differences may have stemmed from time differences between when each breast was last fed from. Glucose also showed significant differences in 60% of participants with 50% showing higher glucose in milk from the left breast and 10% with higher glucose in right breast milk.

Appendix C

Various tubes and feeding rates are used in medicine with huge losses in fat and protein reported [4,5]. Since human milk viscosity shows high dependence on shear rate, the apparent wall shear rate should be maximized to ease milk flow. Methods for tube feeding include continuous drip and bolus feed with a syringe over a specified time period. Calculating the apparent wall shear rate ($\dot{\gamma}_a$) for tube feeding of human milk can be approximated by using the equation

$$\dot{\gamma}_a = \frac{4Q}{\pi r^3} \quad (A5)$$

for a Newtonian fluid flow through a circular duct where Q is the volumetric flow rate and r is the radius of the tube [57]. Using tube diameters [5] and feeding rates [48,58] reported in literature in combination with Equation (A5), a table of approximate apparent wall shear rates for bolus feeds is calculated in Table A1.

Table A1. Approximations of $\dot{\gamma}_a$ in Bolus Feeds over 20 min every 2 h.

Tube Inner Diameter (mm)	Infant Weight (g)	Feed Volume (mL kg^{-1} day^{-1})	$\dot{\gamma}_a$ (s^{-1})
0.5	1000	100	565.9
0.5	1000	200	1131.8
0.5	1500	100	848.8
0.5	1500	200	1697.7
3.0	1000	100	2.6
3.0	1000	200	5.2
3.0	1500	100	3.9
3.0	1500	200	7.9

References

1. Eidelman, A.I.; Schanler, R.J.; Johnston, M.; Landers, S.; Noble, L.; Szucs, K.; Viehmann, L. Breastfeeding and the Use of Human Milk. *Pediatrics* **2012**, *129*, e827–e841. [CrossRef]
2. World Health Organization. *Infant and Young Child Feeding: Model Chapter for Textbooks For Medical Students and Allied Health Professionals*; World Health Association: Geneva, Switzerland, 2009.
3. Vieira, A.A.; Soares, F.V.M.; Pimenta, H.P.; Abranches, A.D.; Moreira, M.E.L. Analysis of the influence of pasteurization, freezing/thawing, and offer processes on human milk's macronutrient concentrations. *Early Hum. Dev.* **2011**, *87*, 577–580. [CrossRef] [PubMed]
4. Stocks, R.; Davies, D.; Allen, F.; Sewell, D. Loss of Breast Milk Nutrients during Tube Feeding. *Arch. Dis. Child.* **1985**, *60*, 164–166. [CrossRef]
5. Jarjour, J.; Juarez, A.; Kocak, D.; Liu, N.; Tabata, M.; Hawthorne, K.; Ramos, R.; Abrams, S. A novel approach to improving fat delivery in neonatal enteral feeding. *Nutrients* **2015**, *7*, 5051–5064. [CrossRef] [PubMed]
6. Bateman, G.; Sharp, P. A Study of the Apparent Viscosity of Milk as influenced by some Physical Factors. *J. Agric. Res.* **1928**, *36*, 647–674.
7. Bienvenue, A.; Jiménez-Flores, R.; Singh, H. Rheological properties of concentrated skim milk: Importance of soluble minerals in the changes in viscosity during storage. *J. Dairy Sci.* **2003**, *86*, 3813–3821. [CrossRef]
8. Trinh, B.; Haisman, D.; Trinh, K.T. Rheological characterisation of age thickening with special reference to milk concentrates. *J. Dairy Res.* **2007**, *74*, 106–115. [CrossRef]
9. Vélez-Ruiz, J.; Barbosa-Cánovas, G. Rheological properties of concentrated milk as a function of concentration, temperature and storage time. *J. Food Eng.* **1998**, *35*, 177–190. [CrossRef]
10. Snoeren, T.; Damman, A.; Klok, H. *The Viscosity of Skim-Milk Concentrate*; Zuivelzicht: Ede, The Netherlands, 1981.
11. Jebson, R.S.; Chen, H. Performances of falling film evaporators on whole milk and a comparison with performance on skim milk. *J. Dairy Res.* **1997**, *64*, 57–67. [CrossRef]
12. Bakshi, A.; Smith, D. Effect of fat content and temperature on viscosity in relation to pumping requirements of fluid milk products. *J. Dairy Sci.* **1984**, *67*, 1157–1160. [CrossRef]
13. Phipps, L. The interrelationship of the viscosity, fat content and temperature of cream between 40 and 80 °C. *J. Dairy Res.* **1969**, *36*, 417–426. [CrossRef]
14. Whitaker, R.; Sherman, J.; Sharp, P.F. Effect of temperature on the viscosity of skimmilk. *J. Dairy Sci.* **1927**, *10*, 361–371. [CrossRef]
15. Magee, H.E.; Harvey, D. Studies on the Effect of Heat on Milk: Some Physico-Chemical Changes induced in Milk by Heat. *Biochem. J.* **1926**, *20*, 873.
16. Evenson, O.L.; Ferris, L.W. The viscosity of natural and remade milk. *J. Dairy Sci.* **1924**, *7*, 174–188. [CrossRef]
17. Zhao, D.b.; Bai, Y.h.; Niu, Y.w. Composition and characteristics of Chinese Bactrian camel milk. *Small Ruminant Res.* **2015**, *127*, 58–67. [CrossRef]

18. Macy, I.G.; Kelly, H.J.; Sloan, R.E. *The Composition of Milks. A Compilation of the Comparative Composition and Properties of Human, Cow, and Goat Milk, Colostrum, and Transitional Milk*; National Research Council: Washington, DC, USA, 1953.
19. Blair, G.W.S. The determination of the viscosity of human milks and the prenatal secretions. *Biochem. J.* **1941**, *35*, 267. [CrossRef]
20. Waller, H.; Aschaffenburg, R.; Grant, M.W. The viscosity, protein distribution, and 'gold number' of the antenatal and postnatal secretions of the human mammary gland. *Biochem. J.* **1941**, *35*, 272. [CrossRef]
21. de Almeida, M.B.d.M.; de Almeida, J.A.G.; Moreira, M.E.L.; Novak, F.R. Adequacy of human milk viscosity to respond to infants with dysphagia: Experimental study. *J. Appl. Oral Sci.* **2011**, *19*, 554–559. [CrossRef]
22. Fondaco, D.; AlHasawi, F.; Lan, Y.; Ben-Elazar, S.; Connolly, K.; Rogers, M. Biophysical Aspects of Lipid Digestion in Human Breast Milk and Similac^{TM} Infant Formulas. *Food Biophys.* **2015**, *10*, 282–291. [CrossRef]
23. Frazier, J.; Chestnut, A.H.; Jackson, A.; Barbon, C.E.; Steele, C.M.; Pickler, L. Understanding the viscosity of liquids used in infant dysphagia management. *Dysphagia* **2016**, *31*, 672–679. [CrossRef]
24. Laogun, A. Effect of temperature on the RF dielectric properties of human breast milk. *Phys. Med. Biol.* **1986**, *31*, 893. [CrossRef]
25. Laogun, A. Dielectric properties of mammalian breast milk at radiofrequencies. *Phys. Med. Biol.* **1986**, *31*, 555. [CrossRef]
26. McDaniel, M.; Barker, E.; Lederer, C. Sensory characterization of human milk. *J. Dairy Sci.* **1989**, *72*, 1149–1158. [CrossRef]
27. Bransburg-Zabary, S.; Virozub, A.; Mimouni, F.B. Human milk warming temperatures using a simulation of currently available storage and warming methods. *PLoS ONE* **2015**, *10*, e0128806. [CrossRef]
28. Dumm, M.; Hamms, M.; Sutton, J.; Ryan-Wenger, N. NICU breast milk warming practices and the physiological effects of breast milk feeding temperatures on preterm infants. *Adv. Neonat. Care* **2013**, *13*, 279–287. [CrossRef]
29. Alatalo, D.; Hassanipour, F. An Experimental Study on Human Milk Viscosity. In Proceedings of the ASME 2016 International Mechanical Engineering Congress and Exposition, Phoenix, AZ, USA, 11–17 November 2016.
30. Mezger, T.G. *The Rheology Handbook: For Users of Rotational and Oscillatory Rheometers*, 4th ed.; Vincentz Network GmbH & Co. KG: Hanover, Germany, 2014.
31. Cadwell, K.; Turner-Maffei, C. *The Lactation Counselor Certificate Training Course Notebook*, 2017–2018 ed.; Healthy Children Project, Inc.: East Sandwich, MA, USA, 2017.
32. Fernandes, R.R.; Andrade, D.E.; Franco, A.T.; Negrão, C.O. The yielding and the linear-to-nonlinear viscoelastic transition of an elastoviscoplastic material. *J. Rheol.* **2017**, *61*, 893–903. [CrossRef]
33. Kline, S.J.; McClintock, F.A. Describing Uncertainties in Single-Sample Experiments. *Mech. Eng.* **1953**, *75*, 3–8.
34. McKennell, R. Cone-plate viscometer. *Anal. Chem.* **1956**, *28*, 1710–1714. [CrossRef]
35. Neville, M.C.; Keller, R.P.; Seacat, J.; Casey, C.E.; Allen, J.C.; Archer, P. Studies on human lactation. I. Within-feed and between-breast variation in selected components of human milk. *Am. J. Clin. Nutr.* **1984**, *40*, 635–646. [CrossRef]
36. Andreas, N.J.; Kampmann, B.; Le-Doare, K.M. Human breast milk: A review on its composition and bioactivity. *Early Human Dev.* **2015**, *91*, 629–635. [CrossRef]
37. Merrill, E.; Gilliland, E.; Cokelet, G.; Shin, H.; Britten, A.; Wells, R., Jr. Rheology of human blood, near and at zero flow: effects of temperature and hematocrit level. *Biophys. J.* **1963**, *3*, 199–213. [CrossRef]
38. Mediwaththe, A.T.M. Impact of Heating and Shearing on Native Milk Proteins in Raw Milk. Ph.D. Thesis, Victoria University, Footscray, Australia, 2017.
39. Elad, D.; Kozlovsky, P.; Blum, O.; Laine, A.F.; Po, M.J.; Botzer, E.; Dollberg, S.; Zelicovich, M.; Sira, L.B. Biomechanics of milk extraction during breast-feeding. *Proc. Natl. Acad. Sci. USA* **2014**, *111*, 5230–5235. [CrossRef]
40. Mortazavi, S.N.; Geddes, D.; Hassanipour, F. Lactation in the Human Breast From a Fluid Dynamics Point of View. *J. Biomech. Eng.* **2017**, *139*, 011009. [CrossRef]
41. Azarnoosh, J.; Hassanipour, F. Fluid-structure interaction modeling of lactating breast. *J. Biomech.* **2020**, *103*, 109640. [CrossRef]

42. Arthur, P.; Hartmann, P.; Smith, M. Measurement of the milk intake of breast-fed infants. *J. Pediatr. Gastr. Nutr.* **1987**, *6*, 758–763. [CrossRef]
43. Daly, S.; Owens, R.A.; Hartmann, P.E. The short-term synthesis and infant-regulated removal of milk in lactating women. *Exp. Physiol. Transl. Integr.* **1993**, *78*, 209–220. [CrossRef]
44. Lucas, A.; Lucas, P.; Baum, J. Pattern of milk flow in breast-fed infants. *Lancet* **1979**, *314*, 57–58. [CrossRef]
45. Lucas, A.; Gibbs, J.; Baum, J. The biology of human drip breast milk. *Early Hum. Dev.* **1978**, *2*, 351–361. [CrossRef]
46. Mitoulas, L.R.; Kent, J.C.; Cox, D.B.; Owens, R.A.; Sherriff, J.L.; Hartmann, P.E. Variation in fat, lactose and protein in human milk over 24h and throughout the first year of lactation. *Br. J. Nutr.* **2002**, *88*, 29–37. [CrossRef]
47. Mizuno, K.; Nishida, Y.; Taki, M.; Murase, M.; Mukai, Y.; Itabashi, K.; Debari, K.; Iiyama, A. Is increased fat content of hindmilk due to the size or the number of milk fat globules? *Int. Breastfeed. J.* **2009**, *4*, 1. [CrossRef]
48. Ogechi, A.A.; William, O.; Fidelia, B.T. Hindmilk and weight gain in preterm very low-birthweight infants. *Pediatr. Int.* **2007**, *49*, 156–160. [CrossRef] [PubMed]
49. Czosnykowska-Łukacka, M.; Królak-Olejnik, B.; Orczyk-Pawiłowicz, M. Breast Milk Macronutrient Components in Prolonged Lactation. *Nutrients* **2018**, *10*, 1893. [CrossRef] [PubMed]
50. Saarela, T.; Kokkonen, J.; Koivisto, M. Macronutrient and energy contents of human milk fractions during the first six months of lactation. *Acta Paediatr.* **2005**, *94*, 1176–1181. [CrossRef] [PubMed]
51. Demmelmair, H.; Koletzko, B. Lipids in human milk. *Best Pract. Res. Clin. Endocrinol. Metab.* **2018**, *32*, 57–68. [CrossRef] [PubMed]
52. Woolridge, M.; Fisher, C. Colic, "overfeeding", and symptoms of lactose malabsorption in the breast-fed baby: A possible artifact of feed management? *Lancet* **1988**, *332*, 382–384. [CrossRef]
53. Apostolidis, A.J.; Armstrong, M.J.; Beris, A.N. Modeling of human blood rheology in transient shear flows. *J. Rheol.* **2015**, *59*, 275–298. [CrossRef]
54. Jensen, R.G. Lipids in human milk. *Lipids* **1999**, *34*, 1243–1271. [CrossRef]
55. Jones, F. *Best Practice for Expressing, Storing and Handling Human Milk in Hospitals, Homes and Child Care Settings*, 4th ed.; Human Milk Banking Association of North America, Inc.: Fort Worth, TX, USA, 2019.
56. Lucas, A.; Gibbs, J.; Lyster, R.; Baum, J. Creamatocrit: simple clinical technique for estimating fat concentration and energy value of human milk. *Br. Med. J.* **1978**, *1*, 1018–1020. [CrossRef]
57. Son, Y. Determination of shear viscosity and shear rate from pressure drop and flow rate relationship in a rectangular channel. *Polymer* **2007**, *48*, 632–637. [CrossRef]
58. Schanler, R.J.; Shulman, R.J.; Lau, C.; Smith, E.; Heitkemper, M.M. Feeding strategies for premature infants: Randomized trial of gastrointestinal priming and tube-feeding method. *Pediatrics* **1999**, *103*, 434–439. [CrossRef]

© 2020 by the authors. Licensee MDPI, Basel, Switzerland. This article is an open access article distributed under the terms and conditions of the Creative Commons Attribution (CC BY) license (http://creativecommons.org/licenses/by/4.0/).

Article

Influence of Oxidation Degree of Graphene Oxide on the Shear Rheology of Poly(ethylene glycol) Suspensions

Yago Chamoun F. Soares [1], Elyff Cargnin [2], Mônica Feijó Naccache [1,*] and Ricardo Jorge E. Andrade [2,*]

1. Department of Mechanical Engineering, Pontificia Universidade Católica-RJ, Rua Marquês de São Vicente 225, Rio de Janeiro 22453-900, Brazil; yago.chamoun@gmail.com
2. Mackgraphe-Graphene and Nano-Material Research Center, Instituto Presbiteriano Mackenzie-SP, Rua da Consolação 896, São Paulo 01302-907, Brazil; elyff.cargnin@mackenzista.com.br
* Correspondence: naccache@puc-rio.br (M.F.N.); ricardo.andrade@mackenzie.br (R.J.E.A.)

Received: 28 February 2020; Accepted: 24 March 2020; Published: 26 March 2020

Abstract: This work studies the influence of the concentration and oxidation degree on the rheological behavior of graphene oxide (GO) nanosheets dispersed on polyethylene glycol (PEG). The rheological characterization was fulfilled in shear flow through rotational rheometry measurements, in steady, transient and oscillatory regimes. Graphene oxide was prepared by chemical exfoliation of graphite using the modified Hummers method. The morphological and structural characteristics originating from the synthesis were analyzed by X-ray diffraction, Raman spectroscopy, thermogravimetric analysis, Fourier transform infrared spectroscopy, and atomic force microscopy. It is shown that higher oxidation times increase the functional groups, which leads to a higher dispersion and exfoliation of GO sheets in the PEG. Moreover, the addition of GO in a PEG solution results in significant growth of the suspension viscosity, and a change of the fluid behavior from Newtonian to pseudoplastic. This effect is related to the concentration and oxidation level of the obtained GO particles. The results obtained aim to contribute towards the understanding of the interactions between the GO and the polymeric liquid matrix, and their influence on the suspension rheological behavior.

Keywords: graphene oxide; polyethylene glycol; rheological characterization

1. Introduction

Graphene was first obtained and characterized in 2004 by two Russian researchers that used the mechanical exfoliation technique to peel a graphite plate and obtain fine graphene flakes [1]. Later, the study of different syntheses of graphene was expanded [2]. Graphene is made up of a two-dimensional network of carbon atoms with sp^2 hybridized, with a very high surface area, and the same hexagonal arrangement as graphite [3–6]. However, it consists of a single flat layer of carbon atoms, while graphite represents stacks of graphene sheets.

Due to its excellent electronic, mechanical, thermal, and optical properties, many studies about graphene have been developed, and its applicability has been increasing in several industrial sectors [7–9]. Although it has excellent properties and numerous applications, there are still some challenges to obtain large quantities of graphene with lower costs and higher structural quality (high purity). This fact has led to the necessity to study and analyze different processes of synthesis [10].

In this work, the synthesis was made by chemical exfoliation. This method is based on the oxidation of graphite, with the addition of oxygenated groups that increase the interplanar distance among the layers, which turns the sp^2 carbons of the lamellae into sp^3, forming the graphite oxide (GrO) [11]. These oxygenated portions introduce a hydrophilic character to the material, making it easier to disperse in water [12]. Using an ultrasound bath, GrO's three-dimensional structure exfoliate into graphene oxide (GO) sheets. Figure 1 shows the schematic model of a graphene oxide sheet formation.

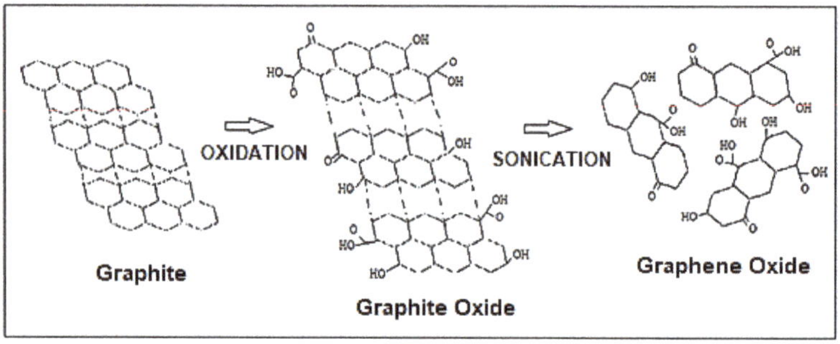

Figure 1. Schematic model of the synthesis to obtain GO by oxidation of graphite.

Graphene oxide has some advantages over graphene, such as better solubility, stability and better dispersion in water and other organic solvents, due to oxidation that turns hydrophobic particles into hydrophilic ones. On the other hand, its electrical conductivity is lower because the functional groups that are inserted contain oxygen, which deactivate double bonds and reduce the conductivity [13–15]. The oxidation process is very important in the exfoliation and dispersion of GO in polymeric matrices. The degree of oxidation is directly connected to the spacing between layers and to the presence of the functional groups developed on the GO sheets [16]. In turn, exfoliation depends on the level of attraction between layers and on the strength of the reaction that occurs between the layers and the solvent.

Since graphene oxide was first obtained, many researches on its thermal and electric properties have been conducted [17]. However, few studies have been carried out on the rheological properties of graphene oxide suspensions. The complete study of the mechanical behavior of these suspensions is necessary to determine their applications and to optimize industrial processes. The addition of nanoparticles to the base fluid, apart from increasing thermal conductivity, may change the rheological, tribological and mechanical properties of the fluid [18–21]. Rudyak (2013) showed that nanofluids may not follow the classical Einstein's relation for viscosity. He observed that the viscosity changes with the particles size [22,23]. Thus, it was possible to conclude that the viscosity of nanofluids depends not only on the surface chemistry, shape or volume concentration of the nanoparticles, but also on their size and concentration [24]. These factors strongly affect the morphology of a suspension of nanoparticles changing the interaction between particles/aggregates due to attractive Van der Waals force [25]. The nanoparticles, as is the case of graphene oxide, can also form clusters with the polymeric chains, which can lead to further increase of the viscosity of the suspension [26].

Naficy et al. (2014) studied the viscoelastic behavior of GO dispersions in water, and concluded that at extremely low concentrations, GO sheets are randomly dispersed in the fluid. As concentration increases, some nematic orders begin to appear, and the storage modulus (G′) increases and exceeds the loss modulus (G″), while the loss modulus remains almost constant with frequency. This can be attributed to the increase in the volumetric fraction of colloidal particles that gives elasticity to the system. An additional increase in concentration results in greater packaging of the nematic phase [27]. Giudice and Shen (2017) also performed the shear rheological characterization of graphene oxide aqueous dispersions. They concluded that a critical concentration exists, below which GO sheets are dispersed, and above which they self-organize. To analyze the responses obtained, the Peclet number (Pe) was evaluated to determine whether the dominant mechanism is Brownian diffusion or convection. In steady-state tests above the critical concentration and at low Pe, GO sheets self-aggregate, while at high Pe, such aggregates dissociate. Moreover, the transient results showed that GO dispersions act as thixotropic fluids [28].

Shu et al. (2015) investigated the effects of graphene oxide and PEG concentrations in pure water on the shear rheological behavior of GO/PEG aqueous dispersion. They concluded that GO aqueous dispersions changed from Newtonian behavior to pseudoplastic as GO concentration increases, and that the critical concentration of isotropic-nematic phase transition was around 6 mg/mL. For the GO/PEG aqueous dispersions, they stated that for all concentrations tested, the viscosity of dispersions decreases with increasing shear rate, and as long as the PEG concentration increases, the viscosity value firstly decreases, and then increases again. This can be explained by the fact that the pure concentrated aqueous GO dispersion is in the liquid crystal phase and, when adding PEG, its chains are adsorbed into the GO sheets, decreasing the viscosity of the aqueous dispersions of GO/PEG. This phenomenon suggests that there is an attractive interaction in the GO/PEG dispersion. Increasing the concentration of PEG increases the strength of the attraction interaction, thereby increasing viscosity [29].

This work aims to study the rheological behavior of a suspension of graphene oxide nanoparticles in a polymeric matrix. Most studies in the literature present the rheological behavior of graphene/polymer nanocomposites focusing on graphene processed from graphene oxide (GO) dispersed in some common solvents such as water. However, there are few reports that aimed the study of obtaining graphene oxide dispersed directly in a liquid polymeric matrix, as will be the focus of this work. It is important to refer as well that some works have analyzed GO suspended in water and GO aqueous suspension in polyethylene glycol (PEG). However, to the best of authors knowledge, none of them addressed the rheological behavior of GO suspended with different degrees of oxidation. Taking that into account, this work intends to analyze the shear rheological behavior of GO suspended in PEG with different concentrations and levels of oxidation.

2. Materials and Methods

2.1. Materials

The products and reagents used for the graphene oxide production were: Graphite powder < 45 µm > 99.99%, Sulfuric Acid (H_2SO_4) 99%, Potassium Permanganate ($KMnO_4$) and Polyethylene Glycol M_w 400 g/mol (all provided by Sigma-Aldrich); Ethanol (C_2H_5OH), hydrochloric Acid (HCl, ACS reagent 37%) and hydrogen peroxide (H_2O_2, 30% m/m) (provided by Synth). The deionized water was supplied by PUC-Rio.

2.2. Preparation of Graphene Oxide Suspensions

The graphene oxide used in this work was synthesized by the modified Hummers method [30]. Several samples with different concentrations of GO were prepared. In a 500 mL round bottom flask, 1.0 g of graphite (Gr) powder and 60mL of concentrated sulfuric acid were added. This reaction is performed under low temperature (emerged in an ice bath) and under stirring at a frequency of 500 Hz for 15 min by a medium goldfish magnetic stirrer. Then, 3.5 mg of potassium permanganate ($KMnO_4$) was slowly added for 15 min, since this reaction is extremely exothermic with a very high kinetic velocity. The purpose of adding these reagents is to expand the structure and oxidize the surface of the graphite sheet due to the presence of oxygenated functional groups. In the end, the system was removed from the ice bath and stirred until the desired oxidation time. The oxidation times used in this work were equal to 2 h and 96 h.

After graphite oxidation, 200 mL of deionized water was slowly added, and the system returned to the ice bath. Then, to stop the reaction, aqueous hydrogen peroxide solution was gradually added until the reaction stopped bubbling so that the flask could be removed from the ice bath and allowed to stand for 12 h.

After decantation the supernatant was discarded and, using a vacuum pump to accelerate the filtration process, the product obtained was washed with deionized water to remove the salts that were still present; aqueous hydrochloric acid solution was used to remove the metal ions; and ethanol to remove the organic residues. To obtain the graphite oxide (GrO) with 2 h of oxidation, the washing process took on average one day, while for obtaining the GrO with 96 h of oxidation, this process took an average of 3 weeks.

Before the product was macerated, it was placed in the vacuum oven for a period of 12 h to remove any remaining humidity. With the intention of preparing the colloidal graphene oxide (GO) suspensions with concentrations ranging from 0.1 mg/mL to 80 mg/mL, as shown in Table 1, the macerated graphite oxide was first dispersed in polyethylene glycol and then exfoliated in ultrasonic bath for a period of 4 h, for both levels of oxidation.

Table 1. Concentration of the PEG-GO suspensions used for the rheological studies.

Concentration mg/mL	Concentration wt.%
0.1	0.01
1	0.1
10	1
20	2
40	4
80	8

2.3. Characterization of Graphite, Graphite Oxide (GrO) and Graphene Oxide (GO)

Several tests were performed to obtain the characterization of graphite, GrO, and GO. The Rigaku MiniFlex II diffractometer determined the material diffraction by the X-ray diffraction (XRD) technique, with the samples powdered at room temperature, using $\lambda CuK\alpha$ and monochromator radiation, varying the scanning at angles of 5° to 50°, with a rate of 2° per minute and fixed slots with 30 kV, 15 mA.

The Alpha 300 R Raman spectrometer (Witec) was used to evaluate the hybridization state of carbon. It was calibrated with a silicon (Si) waffer, integration time of 0.25 s, 50× objective lens and 532 nm laser, and with a grating refraction index of 600 g/mm (grades per millimeters). The sample was obtained by diluting 1 mg of GO in 1 mL of deionized H_2O dropping one drop of the solution onto the silicon oxide substrate.

The SDT-Q600 (TA Instrument) was used for the thermogravimetric analysis of graphite oxide powder. The loss of mass is measured as a function of temperature, while the sample is heated from 25 °C to 1000 °C, with a rate of 10 °C per minute under O_2 atmosphere.

The Bruker Vertex 70 Fourier Transform Infrared Spectrometer was used to obtain the absorption spectrum (FTIR), recorded in the wavenumber range of 750–3500 cm^{-1}. The sample was obtained from dilute solutions of graphene oxides prepared with a concentration of 1 mg/mL in deionized H_2O, and pH > 5. The solution was dripped on a silicon oxide (SiO_2) wafer and after dried the samples were analyzed.

The sheets dimensions and thickness of the GO were obtained by Atomic Force Microscopy, AFM, using the Bruker Model Icon Dimension Microscope with probe, with the tip covered with Si. A resolution of 512 lines with 512 points was captured with Scan Asyst mode in the areas of each image. The samples were prepared from the dilution of 1mg/mL to 0.005 mg/mL of deionized H_2O and previously sonicated for 4 h. It was not possible to perform the AFM for the graphite, because its particles are in the order of 45 micrometers, which is above the upper limit of the microscope, equal to 3 micrometers.

2.4. Rheological Measurements

The rotational rheometer Physica MCR501 (Anton-Paar) with cone-plate geometry was used to analyze the rheological behavior of the suspensions with different GO concentrations. The cone-plate geometry consists of a conical shaped body and a planar circular plate (60 mm diameter and 0.057 mm gap). The angle of the conical body is equal to 6°. The double-gap geometry was also used to obtain comparative data, but in the Physica MCR301 rotational rheometer (Anton-Paar). MCR501 and MCR301 are stress-controlled rheometer. A protective cap is used above the geometry during the tests to avoid solvent evaporation. Each test was repeated at least three times to confirm its reproducibility.

For fluid homogenization, the suspensions were kept under stirring on a magnetic stirrer for 15 min with a constant frequency of 1000 Hz before starting any rheological measurement.

Prior to the measurement, the temperature equilibrium of the samples was set up at 20 °C for 5 min. The flow curves of the PEG suspensions were performed by ranging the shear rate from 1000 s^{-1} to 0.1 s^{-1}. The flow curve was evaluated decreasing the shear rates from the highest to the lowest, since the time to reach the steady state is lower [31]. In order to verify the hysteresis of the curves, tests were also performed, starting from the lowest to the highest rates.

The stability of the suspensions was evaluated as follows: first the flow curve was obtained for all concentrations without shaking them, and then the samples were agitated, and new flow curves were generated and compared to the previous ones.

Dynamic strain sweeps were performed at a fixed frequency of 1 Hz ranging the strain from 10% to 10,000%, to determine the linear viscoelastic region (LVR) of the suspension and to verify the predominance of viscous or elastic effects. Moreover, dynamic frequency sweeps were conducted from 0.1 Hz to 100 Hz, to investigate the structure of the suspensions at a constant shear stress value obtained at LVR.

3. Results and Discussion

3.1. Structural Characterization of Graphene Oxide

Figure 2 shows the XRD patterns of primitive graphite and graphite oxide (GrO) of 2 h and 96 h of oxidation. The pristine graphite presents characteristic peaks at the 2θ = 26.2° relative to the (002) plane. Upon its oxidation, the XRD pattern for GrO displays a new shifted peak at 2θ = 9.90° and 2θ = 9.756° for GrO 2 h and GrO 96 h, respectively, indicating an increase in the d-spacing. The values of 2θ and d-spacing for graphite and graphite oxide layers were achieved by fitting the curves according to Bragg's Law [32]. It can be observed that the d-spacing increase from 0.312 nm (Gr) to 0.895 nm (GrO 2 h) and 0.907 nm (GrO 96 h) due to the presence of oxygen functional groups, such as −OH and −COOH, obtained from the oxidation process [33]. The higher d-spacing obtained for the GrO 96 h is due to the higher oxidation time, which led to an increase in functional groups developed in the material structure [34].

As shown in Figure 3, the Raman spectrum can provide several bands, but the most important ones for carrying out the study and characterization of the load are the D, G, and 2D bands. The 2D band can only be observed in the graphite spectrum between the range of 2500–2750 cm^{-1}. It is related to the two-dimensional structure of the material, providing information about the existence of multilayers that form the graphite [35]. For GO 2 h and 96 h, the Raman spectrum shows no peaks in the 2D region, since it is a defective structure consisting of functional groups intercalated between the graphitic layers. With respect to the band D that is in the range between 1300 cm^{-1} and 1500 cm^{-1}, it is possible to observe a growth in the band D relative to the oxidation process that led to the existence of incomplete bonds, heteroatoms attached to the plane structure and changing the sp^2 to sp^3 bond type on carbon atoms, thus with defects on the edges and layers of the graphite [36]. Comparing the spectra of the samples with respect to the G band, which is between 1500 and 1750 cm^{-1}, there is a widening of the G band, indicating greater disorder in the graphitic structure and greater heterogeneity due to the oxidation process [37]. The I_D/I_G intensity ratio of the D and G bands allows a quantitively estimation of the proportion of defect concentration, depending on the GO oxidation. For the graphite, I_D/I_G = 0.573, the GO 2 h oxidation, I_D/I_G = 1.066 and GO 96 h, I_D/I_G = 1.109, indicating a shift to higher of the ration with the increase of oxidation time. This suggests the size reduction of the in-plane sp^2 domains as the oxidation time increases, and consequently, there was an increase of the structural defects in the carbonaceous structure due to the oxygenated groups and edges on GO 96 h. This observation corroborates the previous XRD results.

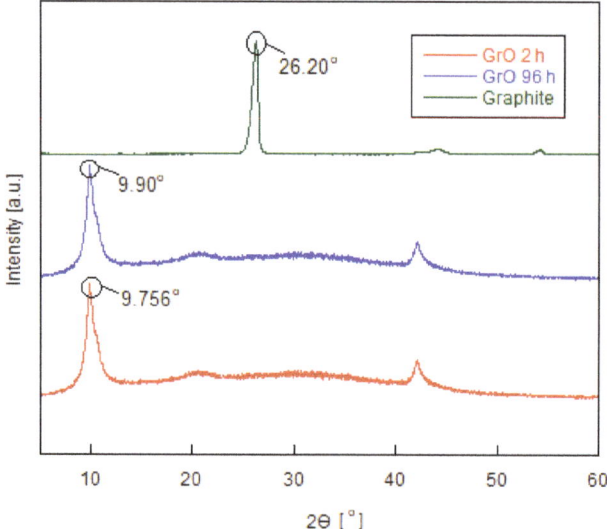

Figure 2. XRD patterns for Graphite, GrO 2 h and GrO 96 h.

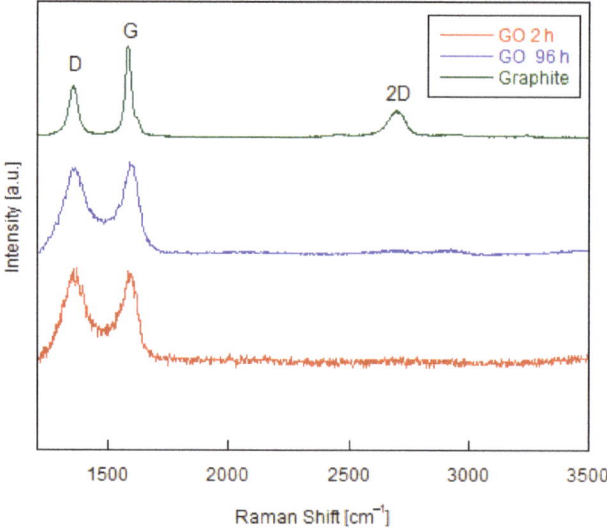

Figure 3. Raman spectrum of Graphite, GO 2 h and GO 96 h.

Figure 4 shows the thermogravimetric curves of graphite, GrO 2 h and GrO 96 h under O_2 atmosphere. For graphite it can be observed that there is thermal stability until 731 °C, when a typical thermal decomposition process of the graphitic material begins to occur and continues until 893 °C. GrO 2 h presents a small loss of mass up to the temperature around 100 °C, resulting from the loss of residual moisture absorbed by the material during the exfoliation process. Between 155 and 300 °C, a further sharp drop corresponding to a loss of 26.74% of the material mass is observed and can be related to the loss of groups that were introduced into the material during oxidation. Finally, the sample has a gradual loss of mass between 500 and 600 °C, related to the process of degradation of the graphite structure. Regarding the results for the GrO 96 h, first a loss of mass resulting from the loss of residual moisture absorbed by the material can be observed and then, between 155 and 300 °C there is 33.38% of mass loss, related to the loss of oxygenated groups that were introduced into the material during oxidation. The remainder shows the mass loss of the graphite material. These values indicate that the GrO's are thermally unstable and the system with the highest oxidation time has the largest number of functional groups introduced in the graphite structure. It is worth mentioning that Jiang et al. (2000) showed that graphite, regardless of its size, is thermally stable because of its high structural stability [38].

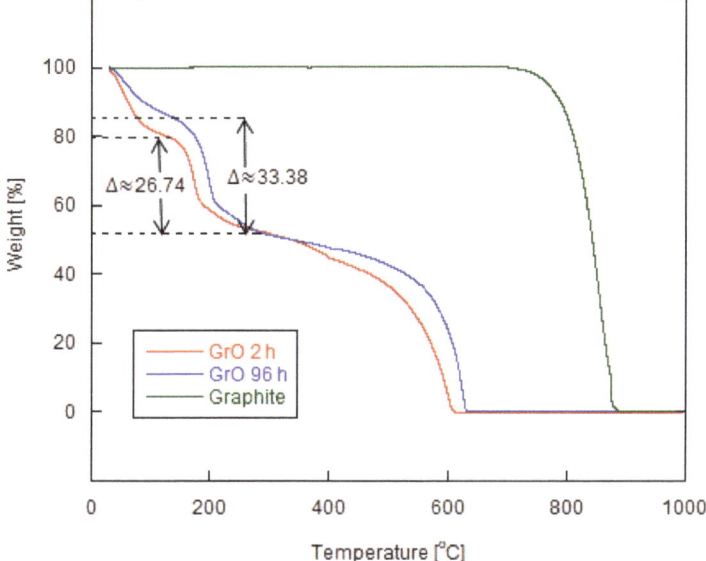

Figure 4. Thermogravimetric curves for Graphite, GrO 2 h and GrO.

Figure 5a,b shows the FTIR spectra of the oxidized GO with 2 h and 96 h. Both results show vibrations regarding the oxygenated functional groups developed by oxidation of the GrO. The peaks at 1055 and 1102 cm^{-1} for the GO 2 h and GO 96 h spectra, respectively, establish the existence of C–O stretching, which can be related to alcohols. The peaks at 1233 and 1277 cm^{-1} also indicate the presence of stretching C–O bonds and can be related to carboxylic acids, esters and ethers (epoxy) [39,40]. The peaks that manifest at frequencies 1422 cm^{-1} for GO 2 h and 1409 cm^{-1} for GO 96 h establish the presence of O–H deformation vibrations. The peaks for the frequencies of 1593 and 1610 cm^{-1} are associated with the C=C stretch. The bands at 1715 and 1711 cm^{-1} determine the stretching of C=O bonds, which may be related to the presence of carboxylic acids and ketones. The broad bands present between the frequencies of 3500 and 2500 cm^{-1} in both spectra are associated with OH in the samples. The peaks at 3340 cm^{-1} for the GO 2 h spectrum and 3211 and 3066 cm^{-1} for the GO 96 h arise from the −OH stretch from carboxylic acids and hydrogen bonds derived from moisture in the samples. However, the peaks that appear at the frequencies of 2920 and 2872 cm^{-1} in the GO 96 h spectrum suggests symmetrical and asymmetric C–H stretching that can be related to the appearance of local defects in the GO nanosheets [41]. The appearance of peaks associated with C–H agree with the results of Raman spectroscopy, which presented a higher I_D/I_G ratio for GO 96 h.

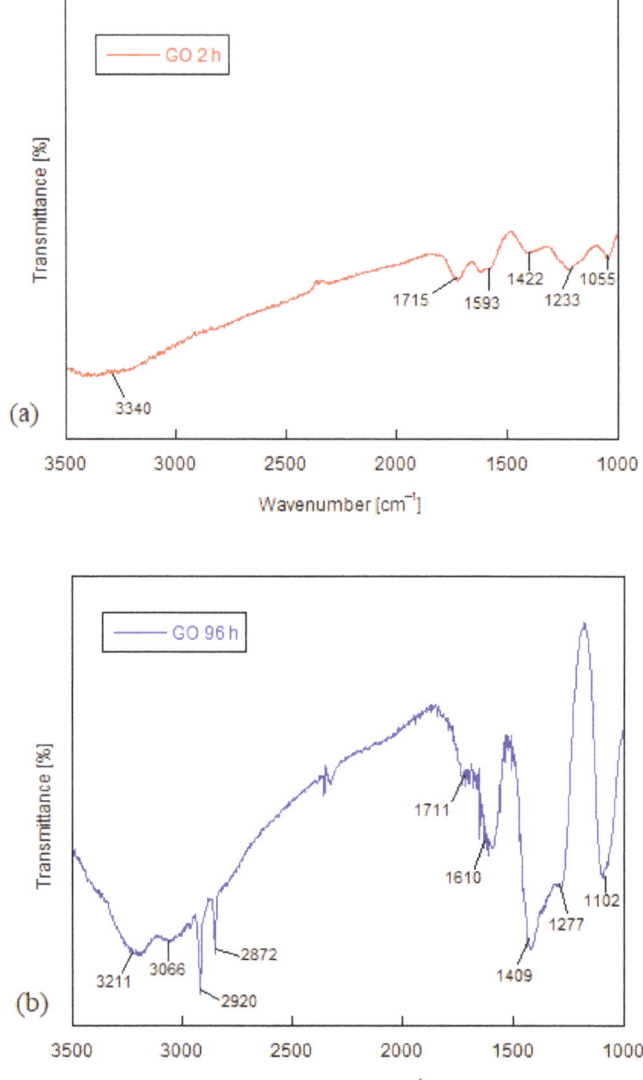

Figure 5. FTIR spectrum of GO with (**a**) 2 h (**b**) 96 h of oxidation.

Atomic force microscopy (AFM) was used to analyze the morphology of the GO obtained, and results are depicted in Figures 6 and 7. Figure 6a presents the results of GO 2 h, and it shows that the particles size range from 60 nm to 294 nm, with more than 50% concentrated between 60 nm and 100 nm, and with an average number of 104 layers, indicating that most of the particles are graphite oxide. Figure 6b shows instead that the GO 96 h has particles ranging from 0 nm to 60 nm, with more than 50% concentrated between 0 nm and 20 nm, and with an average number of 10 layers, in which the GO particles can be classified as multilayer-GO. Figure 7a,b shows a representative image of the GO 2 h and GO 96 h particles respectively, obtained by the AFM in Gwyddion of the exfoliated GO of the aqueous suspension. It is possible to observe, that GO particle 2 h has width between 0.5 µm and 2 µm and a maximum height of 0.11 µm, and for GO 96 h it is seen that the width is between 0.3 µm and 0.8 µm and maximum height of 13 nm. These values are within the results analyzed in Figure 6 and indicates that with the increase of the oxidation degree, it is possible to obtain lower number of layers. However, due to the increase number of defects on the basal plane of the GO 96 h, the layers tend to breakdown more during the sonication process, as indicated by the decrease of the lateral size. These observations are in line with the above characterizations, indicating that with higher oxidation time/degree it was possible to obtain more functional groups in the GO particles, which led to an increase of the exfoliation process.

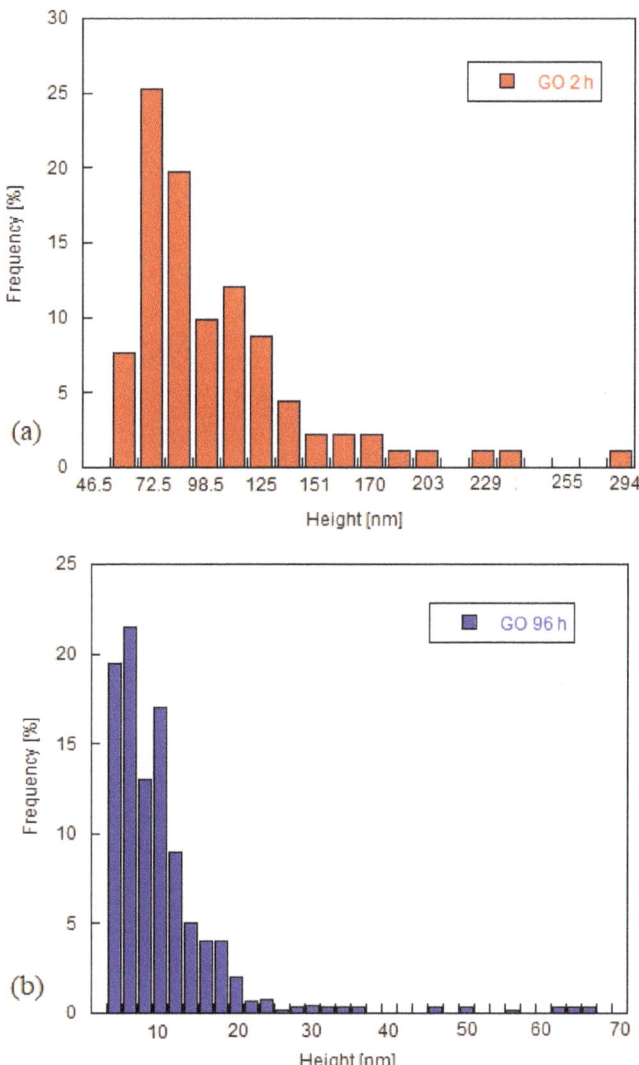

Figure 6. Histograms of frequency (%) as a function of their height (nm) of GO sheets of (**a**) 2 h of oxidation and (**b**) 96 h of oxidation.

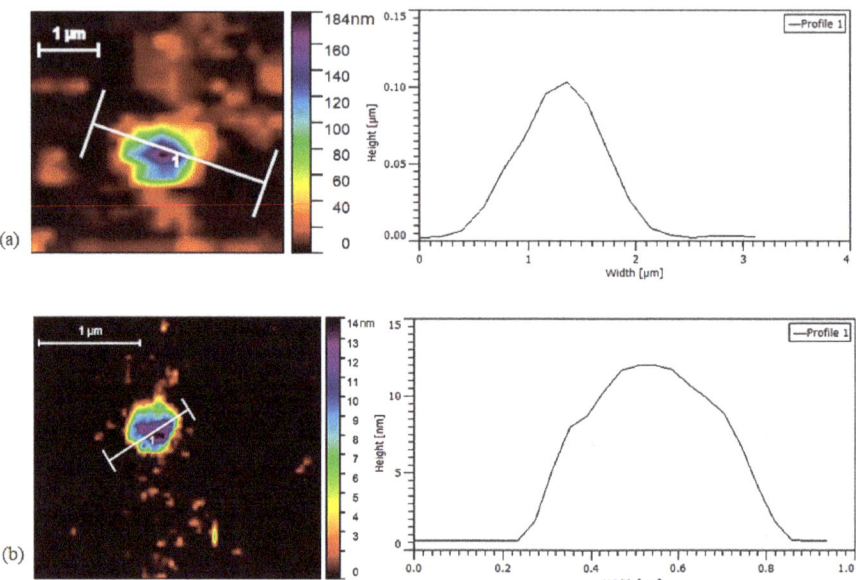

Figure 7. AFM topographic imagens of GO aqueous suspension and the height profile for a single GO sheet of: (**a**) 2 h and (**b**) 96 h of oxidation.

The results obtained with these characterization techniques present similar properties to the GO as in other studies in the literature, showing that the graphene oxide obtained by the modified Hummers method was satisfactory [33]. It was also possible to notice that the GO obtained with the highest oxidation time (96 h) presented fewer layers due to a larger insertion of oxygenated groups between them. This may be justified by the fact that these functional groups further improve the interplanar distance between these sheets and facilitate their separation during the exfoliation process. These oxidized nanoparticles present a hydrophilic behavior, which allows to exfoliate in water.

To evaluate the effect of oxidation level and concentration on the suspension rheology, steady state and oscillatory measurements were performed using rotational rheometry.

3.2. Shear Rheology of Graphene Oxide Suspensions

The difficulty to achieve a stable suspension with no chemical changes such as pH change or surfactant addition, has been a major challenge [42]. To deal with this issue, homogenization of the suspension is carried out by stirring techniques, such as the ultrasonic bath, magnetic stirrer, and high-pressure homogenizer. Factors such as intensity and time of stirring can also influence the suspension behavior.

In order to analyze the stability of the suspensions and to evaluate if the samples have undergone any change due to sedimentation or agglomeration, steady state tests were performed with and without agitation. The stirring was done on a magnetic stirrer for 15 min with a constant frequency of 1000 Hz. Figure 8 shows the flow curves obtained for the graphene oxide suspensions of 2 h of oxidation in PEG with concentrations of 0.1 and 80 mg/mL. It was possible to observe that, at low shear rates, the samples did not demonstrate well-defined rheological behavior while at high rates, when the sample is being stirred more vigorously, the behavior of the suspensions is consistent, mostly due to the orientation/suspension of the GO layers. This demonstrates the fact that these samples need to be agitated before being placed on the rheometer in order to avoid sedimentation.

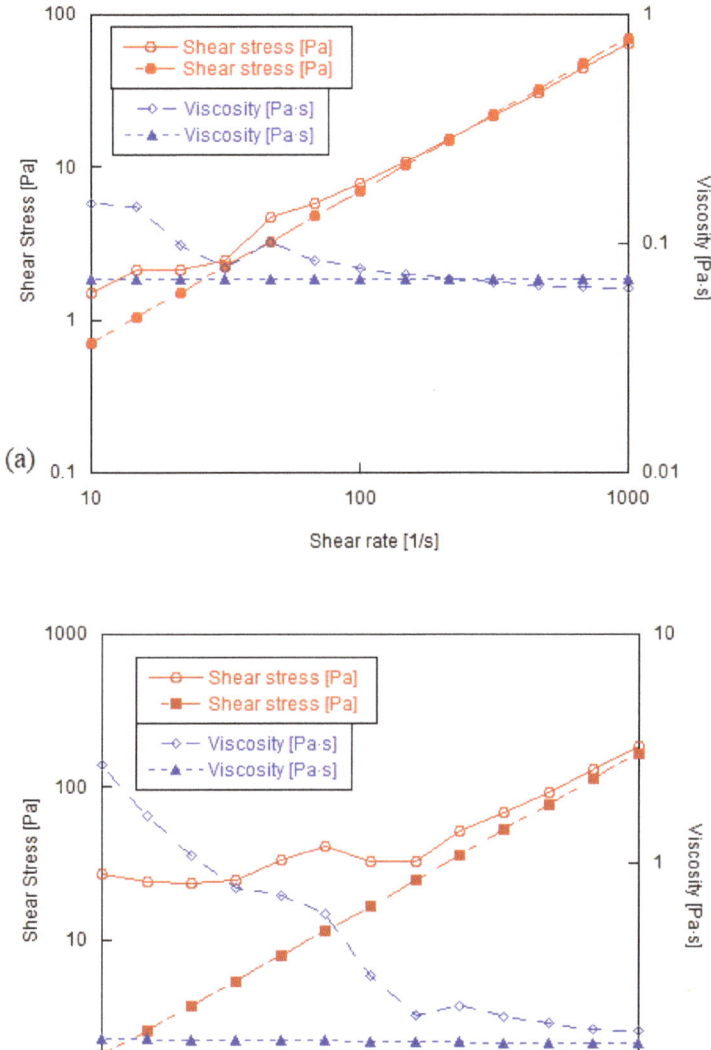

Figure 8. Stability tests for suspensions with concentrations of (**a**) 0.1 and (**b**) 80 mg/mL of GO 2 h comparing suspensions with agitation (filled symbols) and without agitation (open symbols).

Steady state measurements were obtained for each sample applying a shear rate range from $10\ s^{-1}$ to $1000\ s^{-1}$ (without pre-shearing the sample) and then from $1000\ s^{-1}$ to $10\ s^{-1}$. Figure 9 shows that there was no hysteresis in the flow curves generated for 0.1 and 80 mg/mL concentrations made with the GO obtained with 2 h of oxidation. These tests also showed that the execution times of the curves obtained from the highest to the lowest shear rates were lower because the steady state for each point is obtained faster, as expected [31].

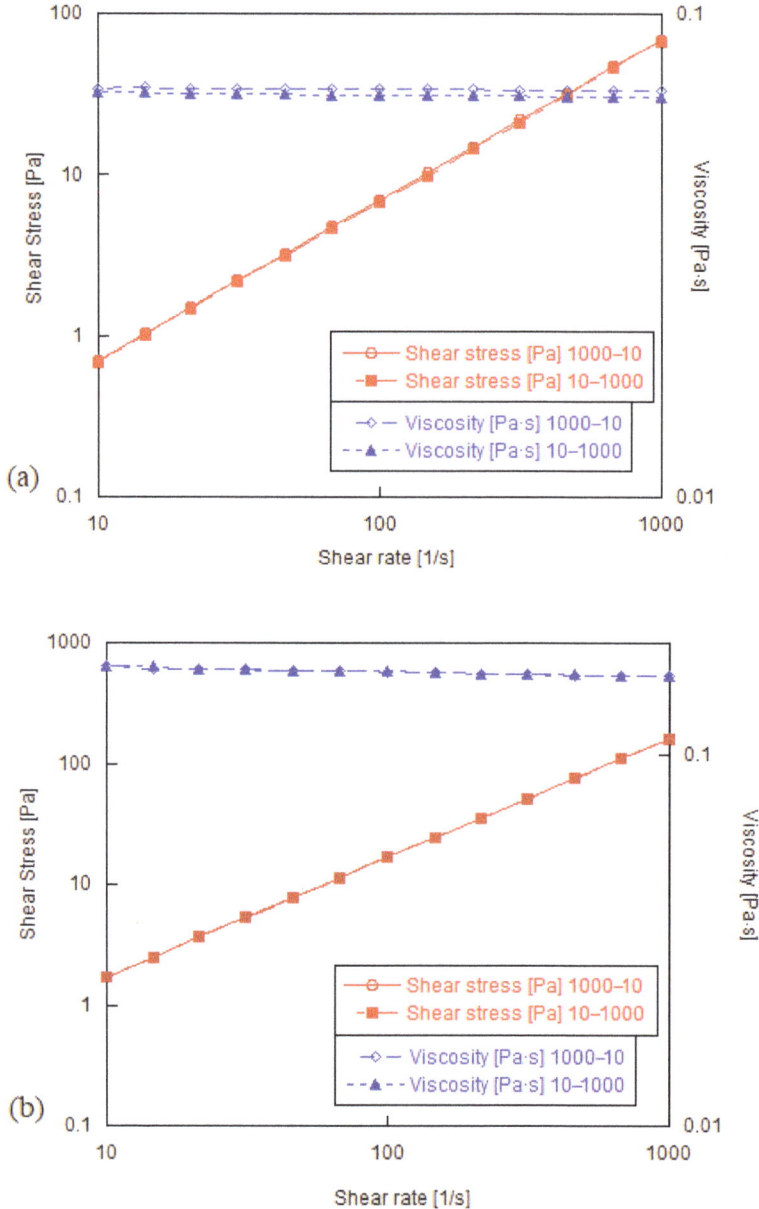

Figure 9. Flow curve applying a shear rate of 10 to 1000 s^{-1} (filled symbols) and then 1000 to 10 s^{-1} (open symbols) for suspensions of GO 2 h with concentrations of (**a**) 0.1 and (**b**) 80 mg/mL.

GO concentration and oxidation degree may influence the rheological properties of graphene oxide suspensions. These effects were investigated for shear rates ranging from 1000 s^{-1} to 0.1 s^{-1} to cover a large variety of possible applications. The steady state curves were obtained for concentrations of GO from 0.1 mg/mL to 80 mg/mL, for oxidation times of 2 h and 96 h. It is important to mention that the low shear rate limit in the flow curves was defined by the low torque limit of the rheometer.

Figure 10 shows the flow curve for suspensions with GO concentration of 40 mg/mL, obtained with oxidation times of 2 h and 96 h. It can be observed that the viscosity of the GO 2 h has a Newtonian behavior, while the GO 96 h presents a slight shear thinning behavior. This suggests the influence of the oxidation level of GO, which also is responsible for the larger viscosity of GO 96 h.

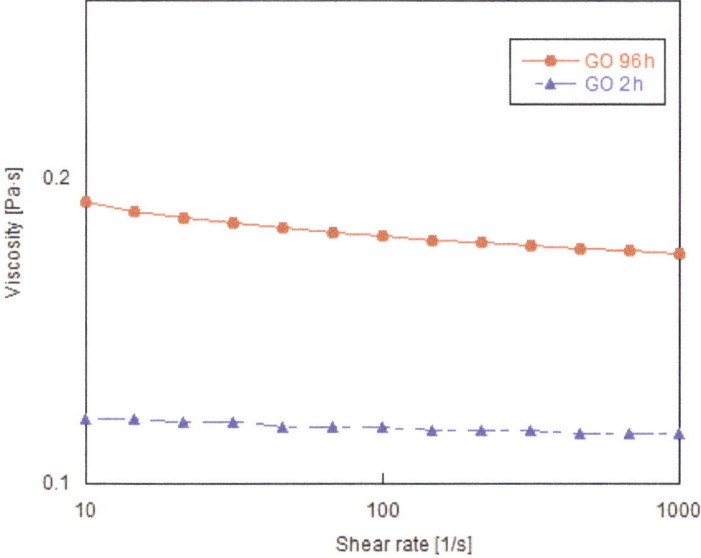

Figure 10. Flow curve comparing viscosity for suspensions with 40 mg/mL of GO with different oxidation times.

The GO particles have an extremely large specific surface and can interact with a large amount of polymer in suspension. Thangavel and Venugopal (2014) evaluated different levels of GO oxidation in aqueous solution and concluded that the more oxidized GO has higher interaction capacity than the less oxidized GO. They argued that the oxidation level of GO increases due to the increase of negatively charged molecules in hydrophilic functional groups [26]. Wang et al. (2004) used the tubeless siphon technique and silica nanoparticles to confirm that one factor that can lead to increasing shear viscosity of a polymer solution is the addition of nanoparticles to the solution, since the extension of polymer bridges between the particles increases the flow resistance [43]. The basic mechanism of this interesting rheological behavior is the Bridging effect [44], which implies that due to the size effect, multiple particles are connected by the intra-chain bridge of a polymer, as it is shown in Figure 11. Therefore, the larger viscosity observed with increasing oxidation time for GO, can be explained by the fact that the longer the oxidation time, higher percentage of oxygenated groups are inserted into the graphite structure, which helps the layers to exfoliated easier, as observed in the AFM and Raman results. This process presents a more dispersed morphology, developing a better interaction between the nanoparticles and the polymer (interaction made by hydrogen bonds), where the PEG chains are connected on the GO sheets forming larger aggregates in the shaped of a network that restricts the suspension mobility and leads to an increase in viscosity. Thus, since the GO obtained with 96h of oxidation was more oxidized as evidenced by the characterization techniques, the viscosity of the suspensions made with this GO is higher than that of the GO obtained with 2 h of oxidation.

Figure 12 shows that for both oxidation times there is an increase in viscosity as the concentration is increased. According to Kamibayashi et al. (2008), increasing particle concentration slowly generates an increase in the suspension viscosity [44]. Einstein's theory of suspension rheology says that the distortion of the velocity field in the surroundings of each particle induces an increase in viscosity [45].

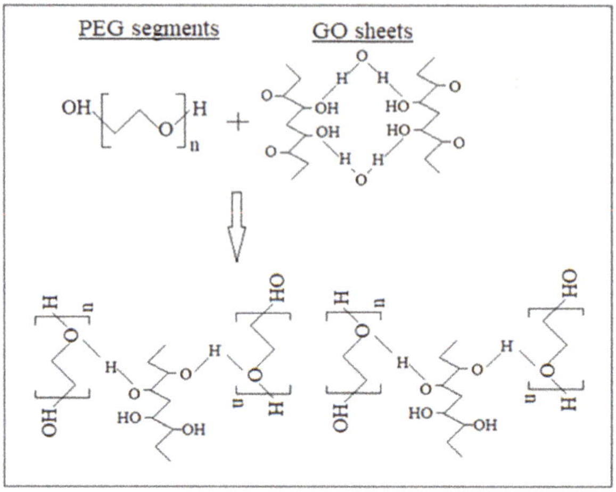

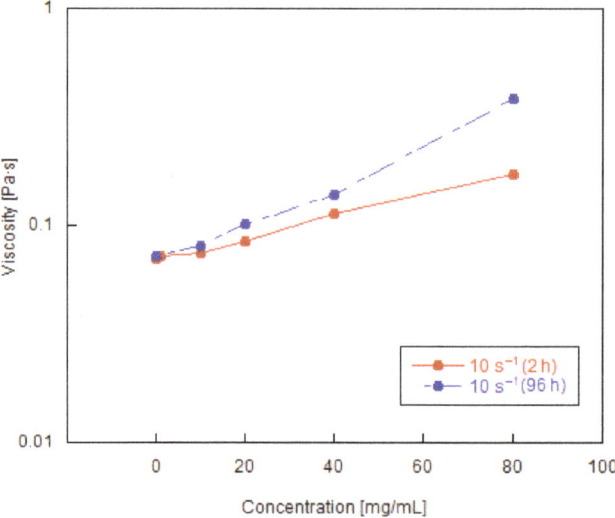

Figure 11. Schematic representation of the Bridging effect.

Figure 12. Shear viscosity of suspensions of GO as a function of different GO concentrations at shear rate $10\ s^{-1}$.

The rheological behavior of these observed suspensions can be discussed in terms of the networks formed between GO nanoparticles and the polymer. As it can be seen from Figure 13, at rest, these networks are kept structured. As a shear rate begins to apply, the network structure is broken, and the particles are oriented along the flow direction. This effect, together with the slip/orientation of the shear GO layers is responsible by the shear thinning behavior that is observed in the more concentrated suspensions [46]. In the lower concentration suspensions, the network is already less structured and therefore no shear thinning is observed. This behavior is shown in Figure 14, which also shows that in the flow curve of the GO 96 h the change of the Newtonian behavior to a pseudoplastic behavior begins to be observed at suspensions with concentration of 40 mg/mL, whereas in the tests made for concentrations of 0.1 to 80 mg/mL of GO 2 h all samples presented a Newtonian behavior in the shear rate range analyzed.

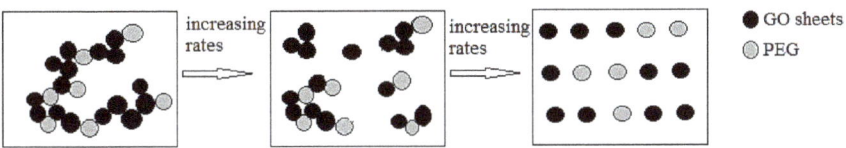

Figure 13. Schematic representation of the breakdown of the structure into flocs with increasing shear rates.

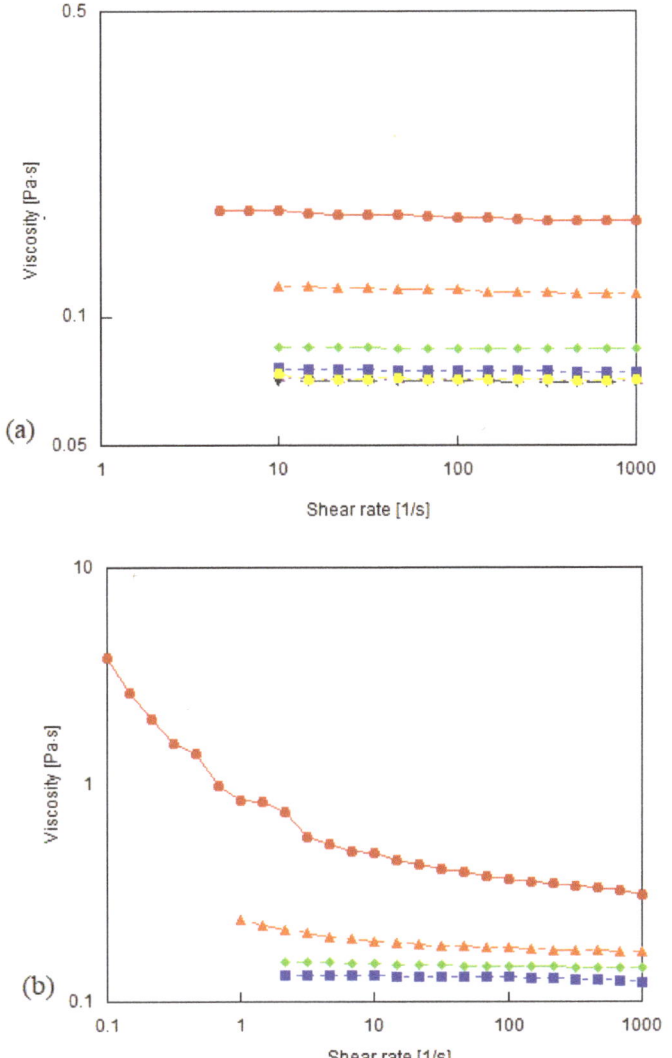

Figure 14. Steady shear rate flow behavior of the (**a**) GO of 2 h (**b**) GO of 96 h of oxidation for concentrations of: ● red—80 mg/mL, ▲ orange—40 mg/mL, ♦ green—20 mg/mL, ■ blue—10 mg/mL, ▶ purple—1 mg/mL, ▼ grey—0.1 mg/mL and ● yellow—0 mg/mL.

According to Shu et al. (2016), in suspensions with very low concentrations of GO, as the sheets of GO are individually dispersed in the polymer with weak interaction among them, the rheological behavior of the GO suspensions is Newtonian. When the GO concentration is above a critical concentration, the molecules have a certain orientational order, exhibiting a characteristic shear-thinning behavior under constant shear flow [29].

Small amplitude oscillatory shear (SAOS) can detect the dynamic mechanical response of complex fluids near equilibrium state. Typical curves of dynamic storage (G') and loss modulus (G") as a function of angular frequency (ω) for the GO suspensions with different concentrations were determined for the GO suspensions. Strain sweep tests were performed in order to obtain the linear viscoelastic region that serves as input data for the frequency sweep tests. The strain used was within the linear viscoelastic range for all tests and the frequency sweep measurement was performed in the range of 0.1–100 Hz for all samples. The low frequency module mainly provides information about the aggregate network and the high frequency response is dominated by the contributions of the polymer matrix [47]. Values below 0.1 Hz have been discarded since they presented a very low torque value, below the equipment limit, and values above 100 Hz are also discarded since the phase angle is not between 0° and 90°, and the material structure might be under irreversible elastic recovery [48].

Figure 15 shows the storage (G') and loss (G") moduli as a function of frequency. For GO 2 h tests, from pure polymer to 40 mg/mL suspension, the loss modulus does not differ significantly, presenting values very close to all concentrations. The storage modulus is practically negligible compared to the loss modulus values, meaning that the material does not present significant elasticity, suggesting poor interaction of GO with PEG. However, the 80 mg/mL suspension began to show slight signs of elasticity. This may be justified by the fact that at low concentrations, GO sheets are randomly dispersed in the polymer matrix and the increase in concentration was too small to cause significant effects on the loss modulus, as the dominant part was still the polymer. For the GO 96 h tests, it can be observed that the suspension with 80 mg/mL GO concentration started to show loss modulus values slightly higher than the suspensions with other concentrations, and the storage modulus started to increase significantly. This may lead to the conclusion that the suspension of GO 96 h with 80 mg/mL concentration is close to the critical concentration, for the present system. These results can suggest that the suspensions presented similar values of the loss modulus, however the higher oxidation time suspension shows a higher storage modulus response, or higher elastic behavior. This observation is another indication that the more oxidized GO, the more exfoliated the layers are, as it was shown above, and consequently a greater interaction with the PEG compared to the less oxidized GO.

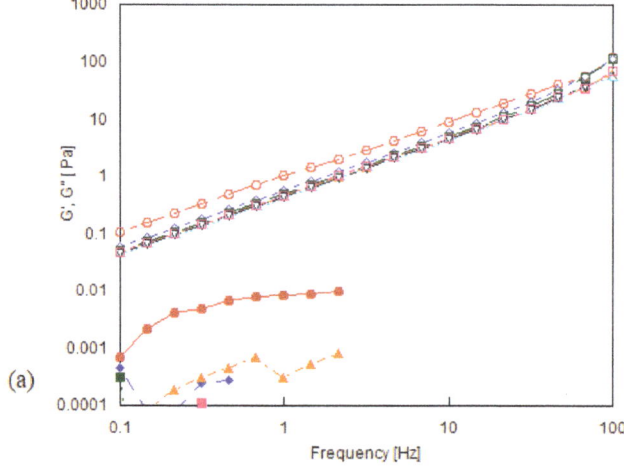

Figure 15. *Cont.*

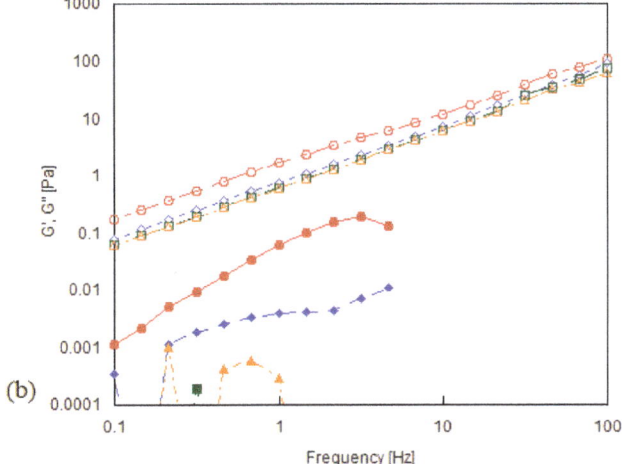

Figure 15. Frequency Sweep of the loss (open symbols) and storage (filled symbols) modulus for suspensions of (**a**) GO of 2 h (**b**) GO of 96 h of oxidation with different concentrations: • red—80 mg/mL, ♦ dark blue—40 mg/mL, ■ green—20 mg/mL, ▲ orange—10 mg/mL, ▶ light blue—1 mg/mL, ▼ grey—0.1 mg/mL and ■ pink—0 mg/mL.

4. Conclusions

This work presented a rheological characterization of suspensions of graphene oxide (GO) nanosheets in polyethylene glycol (PEG). The effect of oxidation time and concentration were investigated. The GO characterization techniques showed that the oxidation occurred and that the longer the oxidation time, the greater the number of functional groups introduced in the graphite structure, increasing interplanar distances. This improves the process of exfoliation, resulting in lower particles size and larger interaction with the polymer chains, due to the bridge effect, which restricts the mobility of the suspension and increases flow resistance. Therefore, larger oxidation times lead to higher suspension viscosity. Regarding the concentration, it was also possible to notice an increase in viscosity with the increase of nanoparticles concentration in the suspensions.

The rheological tests also allowed to investigate the dependence of the suspension viscosity with the shear rate, where it was possible to observe the change from Newtonian to pseudoplastic behavior as the concentration increases, for the GO suspensions prepared with 96 h of oxidation. Moreover, through the oscillatory tests, it was possible to verify that only the most concentrated 96 h GO suspensions showed elasticity.

The results obtained from steady state shear tests and SAOS frequency sweep measurements serve to understand the interaction between nanostructure and the mechanical response of GO/PEG and introduce information about GO/other polymers suspensions. These mechanisms and behavior can provide useful guidelines for nanofluids production.

Since we have demonstrated that the compatibility between PEG and GO has been improved with the degree of oxidation of graphene oxide, this system can be used as well as a model system for studies on electrorheological (ER) suspensions. Such suspensions are important on the field of smart materials, since ER fluids usually consist of semi-conducting or polarizable materials suspended, and their rheological properties can be tuned by electrical field stimuli. Therefore, the introduction of polar functional groups on the surface of the GO, combined with the compatibility with PEG, allow the system to be polarized by an external electric field and subsequently analyzed as a model ER suspension for fundamental studies.

As a suggestion for future work, with the aim to improve the knowledge regarding the properties of graphene oxide suspensions, studies on extensional rheological characterization should be further explored. This type of flow has great importance in industrial processes with large variations in areas, such as flows in extrusion, spray coating, ink jet, among others. In addition, measurements with PEG of other molecular weights in order to evaluate how the bridge effect may interfere with the viscosity of these suspensions are also important.

Author Contributions: Conceptualization, Y.C.F.S., M.F.N. and R.J.E.A.; Formal analysis, Y.C.F.S., E.C., M.F.N. and R.J.E.A.; Funding acquisition, M.F.N. and R.J.E.A.; Investigation, Y.C.F.S. and E.C.; Methodology, Y.C.F.S. and E.C.; Project administration, R.J.E.A.; Resources, M.F.N. and R.J.E.A.; Supervision, M.F.N. and R.J.E.A.; Writing—Original draft, Y.C.F.S., M.F.N. and R.J.E.A.; Writing—Review & editing, M.F.N. and R.J.E.A. All authors have read and agreed to the published version of the manuscript.

Funding: This research received no external funding.

Acknowledgments: The authors would like to acknowledge Fundo Mackenzie de Pesquisa (MackPesquisa, Project Number 181009), Conselho Nacional de Desenvolvimento Científico e Tecnológico (CNPq, Project Number: 409917/2018-4), Coordenação de Aperfeiçoamento de Pessoal de Nível Superior (CAPES), Fundação de Amparo à Pesquisa do Estado do Rio de Janeiro (FAPERJ), Fundação de Amparo à Pesquisa do Estado do São Paulo (FAPESP), and Petrobras. The authors would also to thank Lorena Rodrigues da Costa Moraes for the help provided and fruitful discussion during this study.

Conflicts of Interest: The authors declare no conflict of interest.

References

1. Randviir, E.P.; Brownson, D.A.; Banks, C.E. A decade of graphene research: Production, applications and outlook. *Mater. Today* **2014**, *17*, 426–432. [CrossRef]
2. Neuberger, N.; Adidharma, H.; Fan, M. Graphene: A review of applications in the petroleum industry. *J. Pet. Sci. Eng.* **2018**, *167*, 152–159. [CrossRef]
3. Geim, A.K.; Novoselov, K.S. The rise of graphene. *Nat. Mater.* **2007**, *6*, 183–191. [CrossRef]
4. Konios, D.; Stylianakis, M.M.; Stratakis, E.; Kymakis, E. Dispersion behaviour of graphene oxide and reduced graphene oxide. *J. Colloid Interface Sci.* **2014**, *430*, 108–112. [CrossRef] [PubMed]
5. Ibrahim, A.; Ridha, S.; Amer, A.; Shahari, R.; Ganat, T. Influence of Degree of Dispersion of Noncovalent Functionalized Graphene Nanoplatelets on Rheological Behaviour of Aqueous Drilling Fluids. *Int. J. Chem. Eng.* **2019**, *2019*, 1–11. [CrossRef]
6. Singh, V.; Joung, D.; Zhai, L.; Das, S.; Khondaker, S.I.; Seal, S. Graphene based materials: Past, present and future. *Prog. Mater. Sci.* **2011**, *56*, 1178–1271. [CrossRef]
7. Zhong, Y.; Zhen, Z.; Zhu, H. Graphene: Fundamental research and potential applications. *FlatChem* **2017**, *4*, 20–32. [CrossRef]
8. Yu, W.; Xie, H.; Bao, D. Enhanced thermal conductivities of nanofluids containing graphene oxide nanosheets. *Nanotechnology* **2009**, *21*, 055705. [CrossRef]
9. Frank, I.W.; Tanenbaum, D.M.; Van Der Zande, A.M.; McEuen, P.L. Mechanical properties of suspended graphene sheets. *J. Vac. Sci. Technol. B* **2007**, *25*, 2558–2561. [CrossRef]
10. Lowe, S.E.; Zhong, Y.L. Challenges of Industrial-Scale Graphene Oxide Production. In *Graphene Oxide: Fundamentals and Applications*; Dimiev, A.M., Eigler, S., Eds.; Wiley: Hoboken, NJ, USA, 2017; pp. 410–431.
11. Hongjuan, S.; Tongjiang, P.; Bo, L.; Caifeng, M.; Liming, L.; Quanjun, W.; Xiaoyi, L. Study of oxidation process occurring in natural graphite deposits. *RSC Adv.* **2017**, *7*, 51411–51418. [CrossRef]
12. Zhang, J.; Seyedin, S.; Gu, Z.; Salim, N.; Wang, X.; Razal, J.M. *Liquid Crystals of Graphene Oxide: A Route Towards Solution-Based Processing and Applications*; Wiley: Hoboken, NJ, USA, 2017; Volume 34.
13. Wick, P.; Louw-Gaume, A.E.; Kucki, M.; Krug, H.F.; Kostarelos, K.; Fadeel, B.; Bianco, A. Classification Framework for Graphene-Based Materials. *Angew. Chem. Int. Ed.* **2014**, *53*, 7714–7718. [CrossRef]
14. Imperiali, L.; Liao, K.H.; Clasen, C.; Fransaer, J.; Macosko, C.W.; Vermant, J. Interfacial rheology and structure of tiled graphene oxide sheets. *Langmuir* **2012**, *28*, 7990–8000. [CrossRef]
15. Liu, H.; Zhang, L.; Guo, Y.; Cheng, C.; Yang, L.; Jiang, L.; Zhu, D. Reduction of graphene oxide to highly conductive graphene by Lawesson's reagent and its electrical applications. *J. Mater. Chem. C* **2013**, *1*, 3104–3109. [CrossRef]
16. Shao, G.; Lu, Y.; Wu, F.; Yang, C.; Zeng, F.; Wu, Q. Graphene oxide: The mechanisms of oxidation and exfoliation. *J. Mater. Sci.* **2012**, *47*, 4400–4409. [CrossRef]

17. Hack, R.; Correia, C.H.G.; Zanon, R.A.D.S.; Pezzin, S.H. Characterization of graphene nanosheets obtained by a modified Hummer's method. *Matéria* **2018**, *23*, e-11988. [CrossRef]
18. Aladag, B.; Halelfadl, S.; Doner, N.; Maré, T.; Duret, S.; Estellé, P. Experimental investigations of the viscosity of nanofluids at low temperatures. *Appl. Energy* **2012**, *97*, 876–880. [CrossRef]
19. Ding, Y.; Alias, H.; Wen, D.; Williams, R.A. Heat transfer of aqueous suspensions of carbon nanotubes (CNT nanofluids). *Int. J. Heat Mass Transf.* **2006**, *49*, 240–250. [CrossRef]
20. Wu, Z.; Wang, L.; Sundén, B.; Wadsö, L. Aqueous carbon nanotube nanofluids and their thermal performance in a helical heat exchanger. *Appl. Therm. Eng.* **2016**, *96*, 364–371. [CrossRef]
21. Jama, M.; Singh, T.; Gamaleldin, S.M.; Koc, M.; Samara, A.; Isaifan, R.J.; Atieh, M.A. Critical Review on Nanofluids: Preparation, Characterization, and Applications. *J. Nanomater.* **2016**, *2016*, 1–22. [CrossRef]
22. Rudyak, V.Y. Viscosity of Nanofluids—Why It Is Not Described by the Classical Theories. *Adv. Nanoparticles* **2013**, *2*, 266–279. [CrossRef]
23. Einstein, A. Eine Neue Bestimmung der Molekiildimensionen. *Ann. Der. Phys.* **1906**, *19*, 289–306. [CrossRef]
24. Utomo, A.T.; Poth, H.; Robbins, P.T.; Pacek, A.W. Experimental and theoretical studies of thermal conductivity, viscosity and heat transfer coefficient of titania and alumina nanofluids. *Int. J. Heat Mass Transf.* **2012**, *55*, 7772–7781. [CrossRef]
25. Zhou, Z.; Scales, P.J.; Boger, D.V. Chemical and physical control of the rheology of concentrated metal oxide suspensions. *Chem. Eng. Sci.* **2001**, *56*, 2901–2920. [CrossRef]
26. Thangavel, S.; Venugopal, G. Understanding the adsorption property of graphene-oxide with different degrees of oxidation levels. *Powder Technol.* **2014**, *257*, 141–148. [CrossRef]
27. Naficy, S.; Jalili, R.; Aboutalebi, S.H.; Gorkin, R.A., III; Konstantinov, K.; Innis, P.C.; Wallace, G.G. Graphene oxide dispersions: Tuning rheology to enable fabrication. *Mater. Horiz.* **2014**, *1*, 326–331. [CrossRef]
28. Del Giudice, F.; Shen, A.Q. Shear rheology of graphene oxide dispersions. *Curr. Opin. Chem. Eng.* **2017**, *16*, 23–30. [CrossRef]
29. Shu, R.; Yin, Q.; Xing, H.; Tan, D.; Gan, Y.; Xu, G. Colloidal and rheological behavior of aqueous graphene oxide dispersions in the presence of poly (ethylene glycol). *Colloids Surf. A Physicochem. Eng. Asp.* **2016**, *488*, 154–161. [CrossRef]
30. Sadri, R.; Zangeneh Kamali, K.; Hosseini, M.; Zubir, N.; Kazi, S.N.; Ahmadi, G.; Golsheikh, A.M. Experimental study on thermo-physical and rheological properties of stable and green reduced graphene oxide nanofluids: Hydrothermal assisted technique. *J. Dispers. Sci. Technol.* **2017**, *38*, 1302–1310. [CrossRef]
31. Varges, P.R.; Costa, C.M.; Fonseca, B.S.; Naccache, M.F.; de Souza Mendes, P.R. Rheological characterization of carbopol® dispersions in water and in water/glycerol solutions. *Fluids* **2019**, *4*, 3. [CrossRef]
32. Lopes, C.W.; Penha, F.G.; Braga, R.M.; Melo, D.M.D.A.; Pergher, S.B.; Petkowicz, D.I. Síntese e caracterização de argilas organofílicas contendo diferentes teores do surfactante catiônico brometo de hexadeciltrimetilamônio. *Química Nova* **2011**, *34*, 1152–1156. [CrossRef]
33. De Oliveira, Y.D.; Amurin, L.G.; Valim, F.C.; Fechine, G.J.; Andrade, R.J. The role of physical structure and morphology on the photodegradation behaviour of polypropylene-graphene oxide nanocomposites. *Polymer* **2019**, *176*, 146–158. [CrossRef]
34. Ferreira, E.H.; Andrade, R.J.; Fechine, G.J. The "Superlubricity State" of Carbonaceous Fillers on Polyethylene-Based Composites in a Molten State. *Macromolecules* **2019**, *52*, 9620–9631. [CrossRef]
35. Malard, L.M.; Pimenta, M.A.A.; Dresselhaus, G.; Dresselhaus, M.S. Raman spectroscopy in graphene. *Phys. Rep.* **2009**, *473*, 51–87. [CrossRef]
36. Stankovich, S.; Dikin, D.A.; Piner, R.D.; Kohlhaas, K.A.; Kleinhammes, A.; Jia, Y.; Ruoff, R.S. Synthesis of graphene-based nanosheets via chemical reduction of exfoliated graphite oxide. *Carbon* **2007**, *45*, 1558–1565. [CrossRef]
37. Kudin, K.N.; Ozbas, B.; Schniepp, H.C.; Prud'Homme, R.K.; Aksay, I.A.; Car, R. Raman spectra of graphite oxide and functionalized graphene sheets. *Nano Lett.* **2008**, *8*, 36–41. [CrossRef]
38. Jiang, W.; Nadeau, G.; Zaghib, K.; Kinoshita, K. Thermal analysis of the oxidation of natural graphite—Effect of particle size. *Thermochim. Acta* **2000**, *351*, 85–93. [CrossRef]
39. Socrates, G. *Infrared and Raman Characteristic Group Frequencies: Tables and Charts*; John Wiley Sons: Hoboken, NJ, USA, 2004.
40. Smith, B.C. *Infrared Spectral Interpretation: A Systematic Approach*; CRC Press: Boca Raton, FL, USA, 1998.

41. Luceño-Sánchez, J.A.; Maties, G.; Gonzalez-Arellano, C.; Diez-Pascual, A.M. Synthesis and characterization of graphene oxide derivatives via functionalization reaction with hexamethylene diisocyanate. *Nanomaterials* **2018**, *8*, 870. [CrossRef]
42. Taha-Tijerina, J.J. Thermal Transport and Challenges on Nanofluids Performance. *Microfluid. Nanofluidics* **2018**, *1*, 215–256.
43. Wang, J.; Bai, R.Y.; Joseph, D.D. Nanoparticle-laden tubeless and open siphons. *J. Fluid Mech.* **2004**, *516*, 335–348. [CrossRef]
44. Kamibayashi, M.; Ogura, H.; Otsubo, Y. Shear-thickening flow of nanoparticle suspensions flocculated by polymer bridging. *J. Colloid Interface Sci.* **2008**, *321*, 294–301. [CrossRef]
45. Mueller, S.; Llewellin, E.W.; Mader, H.M. The rheology of suspensions of solid particles. *Proc. R. Soc. A Math. Phys. Eng. Sci.* **2010**, *466*, 1201–1228. [CrossRef]
46. Giannelis, E.P. Polymer-layered silicate nanocomposites: Synthesis, properties and applications. *Appl. Organomet. Chem.* **1998**, *12*, 675–680. [CrossRef]
47. Thomas, S.; Muller, R.; Abraham, J. *Rheology and Processing of Polymer Nanocomposites*; John Wiley & Sons: Hoboken, NJ, USA, 2016.
48. Mezger, T.G. *The Rheology Handbook: For Users of Rotational and Oscillatory Rheometers*; Vincentz Network GmbH & Co KG: Hannover, Germany, 2006.

© 2020 by the authors. Licensee MDPI, Basel, Switzerland. This article is an open access article distributed under the terms and conditions of the Creative Commons Attribution (CC BY) license (http://creativecommons.org/licenses/by/4.0/).

MDPI
St. Alban-Anlage 66
4052 Basel
Switzerland
Tel. +41 61 683 77 34
Fax +41 61 302 89 18
www.mdpi.com

Fluids Editorial Office
E-mail: fluids@mdpi.com
www.mdpi.com/journal/fluids

www.ingramcontent.com/pod-product-compliance
Lightning Source LLC
LaVergne TN
LVHW070154120526
838202LV00013BA/1062